Wolfgang-Gustav Ekkehart Schmidt

Superconducting Nanowire Single-Photon Detectors for Quantum Photonic Integrated Circuits on GaAs

Karlsruher Schriftenreihe zur Supraleitung Band 029

HERAUSGEBER

Prof. Dr.-Ing. M. Noe
Prof. Dr. rer. nat. M. Siegel

Eine Übersicht aller bisher in dieser Schriftenreihe erschienenen Bände finden Sie am Ende des Buches.

Superconducting Nanowire Single-Photon Detectors for Quantum Photonic Integrated Circuits on GaAs

by
Wolfgang-Gustav Ekkehart Schmidt

KIT Scientific Publishing

Karlsruher Institut für Technologie
Institut für Mikro- und Nanoelektronische Systeme

Superconducting Nanowire Single-Photon Detectors
for Quantum Photonic Integrated Circuits on GaAs

Zur Erlangung des akademischen Grades eines Doktor-Ingenieurs
von der KIT-Fakultät für Elektrotechnik und Informationstechnik des
Karlsruher Instituts für Technologie (KIT) genehmigte Dissertation

von M. Sc. Wolfgang-Gustav Ekkehart Schmidt

Tag der mündlichen Prüfung: 16. Juli 2019
Referent: Prof. Dr. rer. nat. habil. M. Siegel
Korreferent: Prof. Dr. W. H. P. Pernice

Impressum

KIT Scientific Publishing

Karlsruher Institut für Technologie (KIT)
KIT Scientific Publishing
Straße am Forum 2
D-76131 Karlsruhe

www.ksp.kit.edu

Print on Demand 2020 – Gedruckt auf FSC-zertifiziertem Papier

ISSN 1869-1765
ISBN 978-3-7315-0980-6
DOI 10.5445/KSP/1000098841

Kurzfassung

Die Entdeckung und das wachsende Verständnis quantenmechanischer Zusammenhänge, das in den letzten Jahrzehnten gewonnen wurde, ermöglicht die Umsetzung einer Reihe neuer Technologien. Quantencomputer können in der Theorie bestimmte Probleme der Informatik signifikant schneller lösen als klassische Computer und quantenmechanische Systeme deutlich effizienter simulieren. Informationstechnische Systeme könnten durch Quantenkryptographie und Quantenschlüsselaustausch deutlich sicherer gemacht werden. In einem quanteninformationstechnischen System ist das Quantenbits (Qubits) die informationstechnisch kleinste Einheit. Die Realisierung von Qubits mit Hilfe von Photonen bietet verschiedene Vorteile: das Photon ist die kleinste Einheit von Quanteninformation und potentiell frei von Dekohärenz, in einem Photon gespeicherte Quanteninformation bleibt typischerweise erhalten.
Die Realisierung von quanteninformationstechnischen Systemen mit photonischen Qubits kann mithilfe eines 2001 durch Knill, Laflamme und Milburn vorgestellten Konzepts erfolgen. Dieses so genannte KLM-Schema erlaubt die Realisierung von Quantenschaltkreisen auf Basis von Elementen der linearen Optik: Einzelphotonenquellen, Strahlteilern und Einzelphotonendetektoren. Die Einzelphotonendetektoren sollten für das KLM-Schema eine Photonenzahlauflösung von vier Photonen besitzen. Analog zu klassischen Schaltkreisen werden komplexe Operationen erst durch das Zusammenschalten einer Vielzahl einzelner Elemente möglich. Um dies zu bewerkstelligen, ist eine Miniaturisierung der einzelnen Elemente erforderlich. Um eine hohe Skalierbarkeit zu erreichen, ist die monolithische Integration aller Elemente auf einem einzelnen Chip notwendig. Dies ist möglich auf Basis eines GaAs-Chips. GaAs erlaubt die Integration von Quantenpunkten als Einzelphotonenquellen und Wellenleitern welche eine Photonenführung und die Umsetzung von Strahlteilern ermöglichen. Supraleitende Nanodraht Einzelphotonendetektoren (SNSPD) können auf Wellenleiter integriert werden und erlauben die Detektion von Einzelphotonen mit einer hohen Detektionseffizienz, einer hohen Zählrate und hoher Zeitauflösung. Um eine hohe Ununterscheidbarkeit von Emittierten Einzelphotonen zu erreichen, was eine notwendige Eigenschaft für die Realisierung eines quantenphotonischen Schaltkreises ist, müssen die Quantenpunkte durch einen zur Energielücke resonanten Laser angeregt werden. Zudem sollten die Wellenleiter nur die Fundamentalmode führen.
Das Ziel der vorliegenden Arbeit ist die Entwicklung von SNSPD für die monolithische, skalierbare Integration aller Kernkomponenten zum Aufbau eines vollständig integrierten quantenphotonischen Chip auf Basis von GaAs.
SNSPD bestehen aus einem Nanodraht mit einer Dicke von typischerweise 5 nm und einer Breite von 100 nm. Um eine vollständige Funktionalität des Detektors bei einer Temperatur von 4.2 K zu gewährleisten, wird NbN als Supraleitermaterial verwendet: Der SNSPD sollte hierzu eine kritische Temperatur von mindestens 10 K besitzen. Die Herstellung von hochqualitativen NbN Schichten auf GaAs stellt eine Herausforderung dar. Eine begrenzte thermische Stabilität des GaAs und die Diffusion von Arsen und Sauerstoff von der Substratoberfläche

in Kombination mit einer starken Gitterfehlanpassung erschweren das Schichtwachstum von NbN auf GaAs. Um eine qualitativ hochwertige Schicht zu erhalten, wurde der Abscheidungsprozess für NbN untersucht. Zusätzlich wurde der Einfluss einer AlN Pufferschicht auf die NbN Schichtqualität untersucht. Mit Hilfe einer 12 nm dicken AlN Pufferschicht konnte eine kritische Temperatur von 12.1 K für eine 5.5 nm dicke NbN Schicht erreicht werden. SNSPDs wurden auf GaAs mit und ohne AlN Pufferschicht hergestellt und miteinander bezüglich ihrer Detektionseigenschaften verglichen. Dafür wurde ein Detektordesign entwickelt das die spätere Integration der Detektoren auf einen phonischen Wellenleiter erlaubt. Das Design basiert auf einer archimedischen Doppelspirale, um den sogenannten „current crowding effect" zu verringern. Durch den Einsatz der Pufferschicht konnte sowohl die Detektionseffizienz gesteigert werden sowie die Dunkelzählrate verringert werden.
Durch den „current-crowding effect" wird die lokale Stromdichte bei kleinen Krümmungsradien auf der Innenseite der Krümmung erhöht, was den kritischen Strom und damit die Detektionseffizienz des SNSPDs verringert und zu einer erhöhten Dunkelzählrate führt. Um den „current-crowding effect" zu verringern, ist ein Spiraldesign nicht für alle Anwendungsfälle die beste Wahl: Auf dünnen Wellenleitern fehlt der notwendige Platz und das Umsetzen eines Detektors mit Photonenzahlauflösung ist nur durch ein komplexes Design mit einer Vielzahl von Schichten zu realisieren. Daher wird in dieser Arbeit das Konzept eines SNSPDs mit einer variablen Schichtdicke vorgestellt. Durch eine deterministische Erhöhung der Schichtdicke soll die Stromdichte in den Knicken verringert werden. Dadurch limitieren die Knicke nicht mehr den kritischen Strom des Detektors. Hierzu wurden Meander-SNSPD mit variabler Schichtdicke und mit gleichmäßiger Schichtdicke hergestellt und verglichen. Durch die variable Dicke konnte eine Erhöhung des kritischen Stroms um $30 \pm 4\,\%$ erreicht werden. Durch die Erhöhung des kritischen Stroms konnte die intrinsische Detektionseffizienz von 32 % auf 97 % für eine Wellenlänge von 500 nm gesteigert werden. Zudem wurde die Dunkelzählrate um eine Größenordnung verringert.
Klassische SNSPDs sind aufgrund ihres zugrundeliegenden Detektionsmechanismus lediglich dazu in der Lage einzelne Photonen zu detektieren. Um die notwendige Mehrphotonendetektion mit SNSPD zu realisieren, wurde ein Konzept umgesetzt, das erstmals 2012 von Jahanmirinejad et al. vorgeschlagen und umgesetzt wurde. Das Design wurde angepasst, um die Integration auf einem photonischen Wellenleiter zu ermöglichen. Die Photonenzahlauflösung wurde für verschiedene Lichtquellen in einem Wellenlängenbereich von 740 nm bis 1550 nm getestet. Die Genauigkeit der gemessenen Photonenstatistik wurde mithilfe von ultrakurzen Laserpulsen evaluiert. Der entwickelte 3-pixel SNSPD auf Saphir bildet die Statistik eines 1550 nm Femtosekundenlasers mit einer Repetitionsrate von 100 MHz bis zu einer mittleren Photonenzahl von 60 Photonen pro Puls korrekt ab. Hierbei erreichte der Detektor eine Detektionseffizienz von 8.5 %. Der entwickelte 4-pixel SNSPD auf GaAs kann die Statistik eines 900 nm Picosekundenlasers mit einer Repetitionsrate von 76 MHz bis zu einer mittleren Photonenzahl von 5 Photonen pro Puls korrekt wiedergeben. Der Detektor erreichte hierbei eine Detektionseffizienz von $22.7 \pm 3.0\,\%$. Um den praktischen Nutzen eines photonenzahlauflösenden Detektors für ein quantenphotonisches System zu testen, wurde die Detektion einer fast ideale InGaAs/GaAs-Einzelphotonenquelle mit der Detektion einer gepulsten Laserquelle mit gleicher Intensität verglichen. Hierbei konnte gezeigt werden, dass ein photonenzahlauflösender SNSPD in der Lage ist. eine klassische Lichtquelle von einer nichtklassischen Lichtquelle zu unterscheiden.

Um einen SNSPD erfolgreich zusammen mit laserangeregten Quantenpunkten und Wellenleitern auf einem GaAs Chip zu betreiben, wurden der Einfluss und die Unterdrückung von unerwünschten Detektionsevents untersucht. Im Vergleich verschiedener Ursachen für unerwünschte Detektionsevents zeigt sich im Falle einer Laseranregung von Quantenpunkten, dass die Mehrzahl unerwünschter Events durch Streulicht des Anregungslasers verursacht werden. Zur Unterdrückung dieses Streulichts wurde der Einfluss von Al/AlN Abdeckungen und eine topographische Veränderung der Chiprückseite untersucht.
Um die Quantenpunktanregung und Detektion der Emission auf einem Chip zu untersuchen, wurden Quantenpunkte, ein einzelner Wellenleiter und ein SNSPD auf demselben Chip implementiert. Die Photonendetektion wurde unter Anregung von Quantenpunktensembles und eines resonant angeregten Quantenpunktes untersucht.
Zur Demonstration aller notwendigen Elemente für quantenphotonische Schaltkreise auf einem Chip auf dem Einzelphotonenniveau, wurde ein vollständiges Hanbury-Brown und Twiss Setup monolithisch integriert. Das Hanbury-Brown Twiss Setup setzt sich aus Wellenleitern mit integrierten Quantenpunkten als Einzelphotonenquellen, einem auf Wellenleitern basierenden Strahlteiler und zwei SNSPDs als Einzelphotonendetektoren zusammen. Durch flächige Al/AlN Abdeckungen auf dem Chip konnte ein Signal zu Rauschverhältnis von 16.2 unter resonanter Einzelphotonenanregung erreicht werden. Die integrierten SNSPDs erreichten hierbei eine Detektionseffizienz von bis zu 47.5 ± 15.7 % für durch den Quantenpunkt emittierte Photonen. Die Wellenleiter führten hierbei nur die fundamentale TE-Mode. Der angeregte Quantenpunkt zeigte eine Feinstrukturaufspaltung und eine Emission mit einem klaren Einzelphotonencharakter mit einer extern gemessenen 2-photonen Korrelation von $g^{(2)}(0) = 0.08 \pm 0.03$ für gepulste resonante Anregung. Dies zeigt eine klare Einzelphotonenemission des integrierten Quantenpunktes. Bei einer TCSPC-Messung mit den integrierten SNSPDs ist ein deutliches Oszillieren zwischen beiden Feinstrukturbestandteilen sichtbar. Das Hanbury-Brown und Twiss Experiment mit den integrierten Elementen wurde unter resonanter Daueranregung und unter gepulst resonanter Anregung des Quantenpunkts durchgeführt. Für Daueranregung wurde ein $g^{(2)}(0) = 0.24 \pm 0.06$ aus den Rohdaten bestimmt. Nach Abzug der Dunkelzählungen wird ein $g^{(2)}(0) = 0.17 \pm 0.04$ erreicht. Dies zeigt, dass über 90 % der gemessenen Photonen aus Einzelphotonenemission stammen. Unter gepulster Resonanzanregung wurde ein $g^{(2)}(0) = 0.59 \pm 0.06$ für die Rohdaten und $g^{(2)}(0) = 0.43 \pm 0.05$ nach Abzug der Dunkelzählungen gemessen. Dies zeigt, dass über 75 % der erfassten Photonen im Falle gepulst resonanter Anregung von der Einzelphotonenemission stammen. Dies zeigt die volle Funktionalität aller integrierten Elemente auf dem Einzelphotonenlevel.
Dies ist die erste erfolgreiche Demonstration eines vollständigen quantenphotonischen Chips auf einer skalierbaren monolithischen Plattform. Das gezeigte Experiment demonstriert erstmalig den kooperativen Betrieb aller Grundelemente für komplexe quantenphotonische Logik auf einem einzigen Chip. Dieses Resultat wurde unter Freiraumanregung bei einer Temperatur von 4 K ohne Verwendung von Time Gating-Techniken und spektraler Filterung erreicht. Dies zeigt das Potenzial von GaAs als Plattform und SNSPDs als Detektoren für Quanteninformationstechnologien.

Table of contents

1 Introduction

At the end of the 19th and the beginning of the 20th century, studies of phenomenons that are caused by ultra small particles like electrons, photons and single atoms started. This led to the observation of effects that did not fit the classical models in physics. In 1900, Max Planck introduced the idea that energy is quantized to derive a formula, Planck's law [1], to describe the observed frequency dependence of radiation emitted by a black body in "an act of desperation". This idea was later used by Albert Einstein to explain the photoelectric effect by the postulation that light can be divided into localized "energy quantas". The idea of describing physical phenomena as quantized is now known as the theory of quantum mechanics and allows the description of natural phenomenons on the smallest scale. The further discovery of quantum effects such as quantum confinement, tunneling or entanglement opened the field for a vast range of new applications. In the field of quantum information science, these applications include among others quantum computation, quantum communication and quantum cryptography. Quantum computers are theoretically able to solve certain problems more efficient than classical computers. Examples for efficiently quantum algorithms are the Grover-algorithm [2], which allows a fast search in large databases, or the Shor-algorithm [3], which allows the factorization of large numbers. The simulation of quantum systems in classical computers suffers from an exponential increase of computing time in dependence on the complexity of the simulated system. This could be solved by quantum computers [4]. In a universal quantum computer [5] the slow down should scale linearly on the complexity, which would result in an exponential speed up. Quantum supremacy is a widely used term to describe the ability of a quantum computer to solve problems that cannot be solved by a classical computer and is expected to be demonstrated in the upcoming years [6]. Quantum communication could allow superadditive scaling which would increase the bandwidth for n given parallel channels by more than n times of the bandwidth of a single channel. Quantum cryptography would significantly increase the security of communication protocols and would allow quantum key distribution which allows the distribution of encryption keys secure against attacks [7,8]. [9]

All of these applications require the implementation of quantum bits or "qubits". In contrast to a normal bit with one state that either represents 0 or 1, a qubit has a fixed pair of reliably distinguishable states that can exist in a continuum of superpositions. To successfully implement quantum computing several requirements need to be met. The qubits need to maintain coherence over the full time of a single operation. The system needs to be scalable to enable complex operations; it should allow the implementation of universal logic and it needs to be correctable to extract the entropy and maintain the computers quantum state to allow the next operation [10]. Several technologies for the implementation of quantum computers are under investigation based on different physical systems with their own advantages and disadvantages. Whilst superconducting qubits have been demonstrated as the most mature technology for the scalable realization of quantum processors [11], they suffer from short coherence times (micro seconds [10]). Nevertheless, the achievement of quantum supremacy using superconducting qubits is expected in the coming years once the error rates have improved [6]. Trapped ions and atoms have very long coherence times (seconds) but require complex trapping and cooling which makes a high scalability difficult [10]. Photon-based quantum systems are very promising. A photon is the smallest unit of quantum information and potentially free from decoherence [12]. Information stored in photons tends to stay there and can be transported over large distances.

1.1 Photonic-Quantum Computing

A scheme proposed by Knill, Laflamme and Milburn (KLM scheme) [13] allows the realization of a complete universal quantum logic by the sole use of linear optical elements. The proposed linear optical quantum computation (LOQC) requires single-photon sources, beam splitters (BS), phase-shifters, single-photon detectors with a four photon resolution and feedback from the detector outputs. A further possibility for quantum computing with the same basic set of photonic elements is boson sampling [14] [15]. It can be implemented with a lower complexity, however, it is not universal. In both approaches photons are used for the implementation of so-called "flying qubits". A single qubit is realized either using polarization entanglement in a single photon by the superposition of two perpendicular polarization states or by a spatial separation using path-entanglement [16].
While first implementations of basic quantum-photonic gates were demonstrated in free space [17] [18], free-space optical systems only have a limited scalability. Complex free-space optical setups are large in size and require extensive adjustments and optimization. This is difficult to accomplish for a large number of elements. The implementation of complex photon-based quantum technologies for quantum metrology [19, 20] and quantum simulators and computers [10, 12, 21] requires the use of a more scalable approach. In classical electronics, the availability of high complex electronic circuits like ASICs and microprocessors was enabled by the invention of integrated circuits by Kilby [22] and Noyce [23] in the 1960s. The use of a scalable monolithic approach, which allows the implementation of all necessary electronic elements onto a single semiconductor, paved the way for a variety of new possibilities in a vast range of applications in automation, computation, metrology and communication. Similarly, an on-chip monolithic integration of a quantum-photonic integrated circuit (QPIC) using a scalable approach, is a major factor in enabling complex applications based on photonic quantum technologies. The demonstration of the feasibility of universal monolithic integration requires the demonstration of all basic elements on a single chip: on-demand single-photon sources, a photonic logic and efficient single-photon detectors.

1.2 Photonic Logic

A photonic logic can be built up by the use of on-chip waveguide(WG)-based photonic elements. Similar to classical integrated circuits (IC) based on transistors, arbitrary logic functions can be implemented by the use of basic quantum-photonic gates in the KLM-scheme. Whilst in classical computing complete logical systems can be built using a basic universal gate like the NAND gate, a complete quantum logic needs the realization of three different elementary gates. Two gates operate on one single qubit each: the Hadamard gate (H) which creates a superposition of both qubit eigenstates and a phase gate which implements a phase shift of $\pi/4$ (T) between the two qubit eigenstates and a third gate operating on two qubits: a controlled-NOT gate (CNOT) [24]. With the utilization of the Hong-Ou-Mandel-effect (HOM) [25], the three elementary quantum gates can be realized as photonic gates by using beam splitters and phase shifters. The used WGs need to operate in a single mode for the used photon wavelength to enable deterministic interference. Using path entanglement, a basic CNOT-gate [26] followed by the implementation of a compiled Shor-algorithm has been demonstrated with full photonic logic on-chip [27]. Waveguide-based elements are scalable and allow for the implementation of a high number of photonic elements [28] [29] [30].

1.3 Single-Photon Detectors

So far WG integration and single-photon detection have been demonstrated for avalanche photodiodes (APDs) [31], transition edge sensors (TESs) [32] and superconducting nanowire single-photon detectors (SNSPDs) [33]. The most investigated detector for QPIC in the recent years are SNSPDs. The successful WG integration of SNSPDs on different material systems has been demonstrated by several groups, in some cases alongside further photonic elements like beam splitters [34] and on-chip photonic filters [35]. A conclusive overview of so far demonstrated WG integrated SNSPDs can be found in [36]. The strong focus on SNSPD in comparison to integrated APDs or TES has several reasons. The integration of the superconducting detectors in a WG is significantly easier than for an APD. The fabrication of an APD requires a complex multilayer process, while the superconducting detector just requires a single additional superconducting layer which allows straight forward WG integration. SNSPD [37] as well as TES [38] can reach detection efficiencies close to unity, whilst InGaAs-APDs have detection efficiencies of up to 50 % [39]. APDs suffer from afterpulsing and have significantly higher dark count rates than their superconducting counterparts. TES require more complex cooling schemes since they usually operate in the milli-kelvin range, while SNSPD can be operated reliably at temperatures as high as 4.2 K, which is in the range of temperatures conventionally used for the reliable operation of quantum dots (QD) single-photon sources. Whilst SNSPDs have no intrinsic photon-number resolution, statistical photon-number resolution can be achieved by the use of additional elements [40] or multiple SNSPDs in combination with beam splitters. SNSPDs (MHz) and APDs (MHz) can reach significantly higher count rates than TESs (kHz) [41]. High efficiencies, low dark count rates and high count rates make SNSPDs the ideal candidate for the fast and reliable operation of a QPIC, since the characterization of optical qubits relies on statistical measurements. In addition, SNSPDs have the best timing accuracy of all known single-photon detectors with timing jitters as low as 3 ps [42]. All these properties make SNSPDs the ideal detector for QPIC and they are consequently the detector of choice.

1.4 Material Systems for the Full Monolithic Integration of a Quantum-Photonic Circuit

Whilst SNSPDs have been demonstrated on a vast range of material systems suitable to build up integrated photonic logical elements [36], the integration of single-photon sources significantly limits the material choice [43]. To enable quantum computing, an integrated photon source needs to meet several requirements. Single photons need to be generated at a high efficiency with a high degree of indistinguishability. To allow a QPIC to operate reliably, the single photons need to be generated on-demand. The source needs to be tunable to allow the generation of indistinguishable single-photons out of two independent sources. Silicon-based material systems like silicon-on-insulator (SOI), silica-on-silicon and silicon-nitride-on-silica, profit from a well-developed technological process developed for CMOS technology which allows the on-chip integration of a significant number of photonic elements [28–30,44]. While silicon-based platforms benefit from a low absorption for photons in the near infrared range, the generation of single photons in this materials is limited to spontaneous down conversion (SPDC) sources [45] and spontaneous four wave mixing (SFWM) sources [46] [47] [48]. SPDC and SFWM sources provide single photons with a high degree of indistinguishability. However, they require intense pump laser sources [49], hence complex filters are required for stray light on-chip in order to avoid noise for integrated detectors. In addition, single photons are generated probabilistic and not on-demand. Lithium niobate ($LiNbO_3$) is a highly versatile optical material with a strong optical non-linearity which allows the integration of high-quality quantum-photonic elements in well-developed technological processes [50]. However, due to a low refractive index, the mode diameter for optical and infrared photons in $LiNbO_3$ is large, which leads to a large size of photonic structures, making the integration of a high number of photonic elements challenging. For Ti in-diffused WGs, which are typically used within $LiNbO_3$, the optical mode is guided deep inside the substrate, which makes an efficient detector integration challenging [51]. Single-photon generation in this material is well-developed for SPDC, in addition, the integration of QDs has been demonstrated recently [52]. The most promising materials for monolithic integration are materials that allow the direct integration of compact two-level systems as deterministic, on-demand single-photon sources [53]. Defect centers in diamond can be used as single-photon sources [54]. These defect centers allow the deterministic emission of single photons with a high degree of indistinguishability for spatially distributed defect centers [55]. However, the fabrication of diamond layers with "optical quality", which are needed for photonic circuits, is challenging [56]. The most advanced platforms for the integration of single-photon sources are III-V semiconductor-based systems, uning quantum dots (QDs). Self assembled InGaAs/GaAs-QDs integrated into GaAs allow the generation of single photons with an efficiency of up to 100 % [57]. This allows an excitation of the QD with low pumping powers. Furthermore, InGaAs/GaAs-QDs can be used as on-demand sources of single photons with near unity indistinguishability [58]. GaAs allows the fabrication of low loss WGs [59] and enables compact devices and circuits due to a tight mode confinement connected to its high refractive index [60]. The large electro-optical effect in GaAs makes the implementation of active controlled phase shifters [61] possible. The emission can be tuned externally using temperature, E-field (Stark-effect) [62] or strain tuning [63] which allows the emission of indistinguishable photons out of two remote QDs [62, 64]. Single-photon emission in combination with single-mode waveguides (WGs) and beam splitters (BS) was demonstrated using above-band [65–67] and resonant excitation [68–70]. The integration of niobium-nitride(NbN)-based SNSPDs as single-photon detectors on GaAs was demonstrated by various groups [33,35,71,72]. This versatility makes GaAs the most promising candidate for a full monolithic integration of QPIC [36, 43, 60]. Approaches hybridizing two material systems have been demonstrated but require complex fabrication techniques making a scaling of elements difficult if not impossible [73–76].

1.5 SNSPDs on GaAs

The SNSPD was first proposed by Semenov et al. [77] and demonstrated by Goltsman et al. [78] in 2001. The detector itself consists typically of an ultra-thin superconducting film at a thickness in the range of $d = 4$ to 6 nm which is structured into nanowires of a width $w \leq \approx 100$ nm. To increase its area, the nanowire is typically structured in a meander shape. The detection mechanism is based on local quenching of the superconductor. Whilst SNSPDs are click/no-click detectors without intrinsic energy and photon-number resolution, they are almost ideal detectors for photonic applications. In an optimal configuration, they can have a detection efficiency close to unity [37], a timing resolution as high as 3 ps [42], a count rate in the upper megahertz regime and do operate in the full optical and near to mid infrared range. The most common superconductor for SNSPD fabrication is NbN. The high critical temperature of NbN of up to 16 K (bulk) allows the SNSPD operation in liquid helium at 4.2 K. While GaAs is a very promising substrate for the full integration of all photonic elements necessary for an integrated photonic chip, the fabrication of high-quality NbN-based SNSPDs is challenging. A large lattice mismatch of about 27 % [79] between GaAs and NbN induces stress and disrupts a uniform growth of the NbN film [80] [81]. Naturally occurring surface oxides in combination with a low temperature stability make the growth of high-quality NbN films on GaAs challenging. At temperatures above 535 °C, which are typically used for the deposition of high-qualitiy NbN films, arsenic and oxygen are desorbed from the GaAs surface. This results in a growth of surface pits, causing a roughened-up surface [82]. Interdiffussion of oxygen and arsenic into the NbN film during film growth leads to the formation of an intermediate layer of Nb-O and Nb-As compounds [80] with destroyed superconductivity. This normal conducting layer being in close proximity to the superconducting layer reduces the effective thickness of the superconducting film and further suppresses the critical temperature. This suppression of superconductivity is known as the superconducting proximity effect [83] [84] [85]. Consequently, the fabrication of high-quality SNSPDs on GaAs requires a careful optimization process.

1.6 Photon-Number Resolving Detectors

As previously discussed, photon number resolving (PNR) single-photon detectors are essential for quantum computing using the KLM-scheme or boson sampling. PNR detectors with 4-photon resolution are necessary for the implementation of a KLM-scheme and are conse-

quently of special interest as detectors for complex integrated quantum-photonic circuits [86]. The important benefit of a PNR detector is the direct characterization of photon-number states [87]. The direct measurement of a photon-number correlation can be used for quantum enhanced metrology [88], for quantum random number generators for quantum cryptography [88] or as receivers for quantum optical communication [89]. Light from a light source is typically characterized by its coherent properties by a series of normalized Glauber functions $g^{(n)}$ [90]. The $g^{(2)}$ value at zero time delay is commonly used to identify the fidelity of a single-photon source [91]. An n-PNR detector that is appropriately fast is able to directly measure the photon-number correlations of a light source to the n-th order. Two 50/50 beam splitters and a delay line in combination with two single-photon detectors with at least 2-photon number resolution are sufficient to characterize the photon-number correlation of a single-photon source to the third order while simultaneously determining its photon indistinguishability [91]. PNR-detectors allow for a reduction of the number of circuit elements in a PIC and consequently a higher integration density. n single-photon detectors with n 50/50 beam splitters may be replaced by a single n-photon-number resolving detector in dependence of the circuit design [91].
Whilst there are hints that the SNSPD rise time may be used to distinguish single from multi-photon events, the detector's rise time needs to be measured with a high accuracy, which is challenging [92] [93]. PNR for SNSPD is reliably achieved by using multiple detector elements, which also allows spatial resolution [94, 95]. Multiple detector elements require an individual readout and biasing line for each pixel or a complex frequency [96, 97] or time-tagged multiplexing scheme [98] which can be combined with a differential readout [99]. Where spatial resolution is not of interest, PNR SNSPDs that utilize amplitude multiplexing are the easiest to read out and operate. Only one readout and biasing line for each detector is needed, so they can be used within a standard SNSPD setup. Amplitude multiplexing can be achieved by dividing the detector into different sections, by placing resistors in series of parallel nanowires [81, 100] or in parallel to series nanowires [40, 101–104], resulting in an output pulse with an amplitude related to the number of triggered elements. Parallel nanowires have the disadvantage that the detection of one photon shifts the working point of the detector for the simultaneous detection of another photon, which limit this type of detector to a dynamic range of 4 to 6 photons and the efficiency to a few percent [100]. Photon-number resolving SNSPDs based on series nanowires, on the other hand, have been shown to be able to operate over a high dynamic range of photons [40, 104]. Furthermore, their waveguide integration has been demonstrated for multiple pixels on one waveguide [105] and on two waveguides with one pixel each [106].

1.7 Full Integration of a Quantum-Photonic Circuit – State of the Art

The simplest demonstration of the successful operation of integrated single-photon sources, WGs and single-photon detectors together on the same chip, is done by the full integration of a Hanbury-Brown and Twiss setup (HBT) [107]. This setup consists of a 50/50 beam splitter and two detectors and allows the demonstration of single-photon operation by a measurement of the two photon correlation $g^{(2)}$ of the light source. For a $g^{(2)}(0) < 0.5$, the system is regarded to be on the single-photon level. So far three groups demonstrated the simultaneous integration of a single-photon source, WGs and single-photon detectors. Reithmaier at al. were the first to demonstrate the monolithic integration of QDs, WGs and NbN SNSPDs based on the GaAs material system in 2013 [80]. However, a multimode WG was used and strong stray light during resonant excitation of the QD made temporal filtering necessary. While they were able to characterize the QD lifetime on-chip, single-photon operation was only demonstrated using spectral filtering off-chip in combination with off-chip detectors [108]. In 2016, Digeronimo et al. were also able to show the integration of QDs, WGs and NbN SNSPDs on a single GaAs chip [35]. Again, a multi-mode WG was used, this time with a WG-integrated spectral filter. The QDs were operated under above-band excitation and no operation on the single-photon level was reported. The first successful demonstration of a fully integrated platform was shown in 2016 by Khasminskaya et al. [109]. By integrating of a carbon nanotube, a single-mode WG and NbN SNSPDs on a silicon nitride platform, they were able to demonstrate the full on-chip-operation on a single-photon level with a $g^{(2)}(0) = 0.49$. However, this approach is non-monolithic, difficult to scale and the observed single-photons stem from electrical excitation, which so far only allows single-photon generation with a limited degree of indistinguishability. In addition, the discussed integrations lack a scalable beam splitter and, therefore, one of the basic elements needed to build up photonic logic.

1.8 Aim of This Work

The goal of this work is the development of SNSPD for the scalable integration of all necessary elements for a quantum-photonic chip on the GaAs platform. To allow for detector operation at a temperature of 4.2 K, NbN is used as a detector material. The growth of high-quality NbN films needs to be optimized on GaAs to allow the implementation of SNSPDs with a $T_C > 10$ K, to allow reliable operation at 4.2 K at an $I_C > 10\,\mu A$, to provide a sufficient signal-to-noise ratio (SNR) for detector pulses. Detectors need to be designed and tested in a design suitable for waveguide integration with a system jitter >100 ps. The design needs to be optimized to operate at a high intrinsic detection efficiency at a wavelength of 900 nm, which is the emission wavelength of the integrated InGaAs/GaAs QDs. SNSPDs with photon-number resolution need to be implemented and demonstrated. The influence and suppression of stray light from the excitation laser for QDs used for on-chip photonic measurements needs to be investigated. The integration of detectors with waveguides and quantum dots need to be demonstrated. Finally, the operation of QD single-photon sources, waveguides, beam splitters and SNSPDs together on a single chip has to be demonstrated on a single-photon level.

1.9 Outline

Chapter 2 All basic elements for a QPIC on GaAs are introduced: passive elements for photonic circuits, QD single-photon sources and SNSPDs. Their physical operation principles are explained with a special focus on SNSPD. The SNSPDs operation is introduced, material

and geometrical limitations are discussed. Concepts for photon-number resolving SNSPDs utilizing amplitude multiplexing are introduced.
Chapter 3 The benefit of an AlN buffer layer for the growth of high-quality NbN films is investigated and the influence of growth parameters for NbN and AlN on the NbN film quality is characterized. The RF-properties of grown AlN films are investigated and used for the implementation of an on-chip high-pass filter.
Chapter 4 The design and fabrication of double-spiral type SNSPDs for WG integration is discussed. The benefit of an AlN buffer layer for SNSPDs on GaAs is investigated. SNSPDs in a design suitable for WG integration are fabricated with and without buffer layer and compared regarding their superconducting and optical properties.
Chapter 5 A SNSPD design using a variable film thickness is proposed to reduce the influence of the current-crowding effect. The design intends to improve the detection efficiency and to reduce the dark count rate of a SNSPD with sharp bends. SNSPDs are fabricated with variable and uniform thickness and compared regarding their superconducting and optical properties.
Chapter 6 PNR-SNSPDs in a series design are developed on sapphire and for WG-integration on GaAs. The presence of photon-number levels is demonstrated for a broad range of wavelengths and light sources. The detector statistics is recorded for ultra-fast pulsed lasers with a pulse rate of 76 MHz and 100 MHz and compared to the expected statistic. The responses of a PNR-SNSPD for a Poissonian and sub-Poissonian light source are compared.
Chapter 7 The design and fabrication of the individual elements needed for a QPIC is introduced. The influence of noise counts on a SNSPDs with a focus on stray-light caused by QD excitation is investigated. The excitation and detection of QD emission using WG integrated SNSPDs is demonstrated. A fully on-chip integrated HBT setup is introduced and all elements are characterized individually. The full operation of the HBT setup is investigated for pulsed resonant and cw resonant excitation.
Chapter 8 The obtained results are summarized and possible optimizations are discussed shortly.

2 Elements for Quantum-Photonic Integrated Circuits

This chapter gives a brief introduction to basic concepts and elements necessary for quantum logic based on quantum-photonic integrated circuits (QPIC). The Hong-Ou-Mandel (HOM) effect and the structure of quantum gates using linear optics are described with their realization using on-chip passive photonic elements. The structure and optical excitation of QD-based single-photon sources is explained and the detection mechanism, as well as the design restrictions for SNSPDs are presented.

2.1 Integrated Passive Photonic Elements

A full photonic logic in the framework of the scheme for linear optical quantum computing (LOQC) presented by Knill, Laflamme and Milburn [13] (KLM-scheme) can be implemented with two basic elements. The first element is a WG to confine photons, guide them on-chip and implement passive phase shifts. The second element is a beam splitter to allow the interaction between photons that constitute an optical qubit. The utilization of the HOM interference is hereby key. The dimensions of couplers and WGs used during this thesis were simulated and developed by U. Rengstl [110].

2.1.1 Hong-Ou-Mandel Effect

The Hong-Ou-Mandel effect (HOM) is a two-photon interference effect first described by Hong, Ou and Mandel [25]. The effect describes the behavior of two indistinguishable photons entering a 50:50 beam splitter (BS). Indistinguishability requires the photons to have identical quantum states, including an identical spectral, temporal and spatial profile and polarization. When a single photon enters an ideal 50:50 BS, it has a probability of 50 % of being reflected and of 50 % of being transmitted. When two photons enter the beam splitter

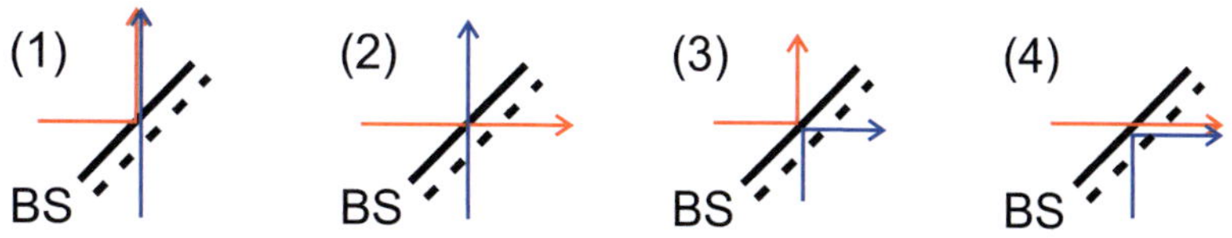

Figure 2.1: Possibilities of reflection and transmission of two photons entering a BS at two opposing sides. The side where the beam splitter introduces a phase shift π is indicated by a dashed line.

at different sides, four possibilities for the behavior of photons exist (fig. 2.1). Either the photon from the top is reflected and the photon from below is transmitted (1); both photons are transmitted (2); both photons are reflected (3); the photon from below is reflected and the photon from the top is transmitted. For (1) and (4) both photons exit the BS on the same side, for (2) and (3) both photons exit the BS on different sides. Reflections from the bottom side of a symmetrical BS introduce a phase shift π, which corresponds to a factor of -1 for the amplitude. For indistinguishable photons possibilities (2) and (3) have the same value but opposite signs due to the introduced phase shift which causes destructive interference and the probability for (2) and (3) go to zero. Consequently, indistinguishable photons must leave the beam splitter at the same output port. In order to have the maximum visibility of the HOM, the following conditions need to be fulfilled: there must never be more than one photon on either input port of the BS, the two photons need to be indistinguishable and both photons need to be in pure states. In case of a QPIC, photons emitted from the integrated photon sources must therefore be indistinguishable and in pure states for a successful operation [16, 110, 111].

2.1.2 Basic Photonic Gates

To have an universal set of gates, T, H and CNOT gates need to be implemented in linear optics. By using these three elementary gates, a Toffoli gate [112] can be implemented that acts as an universal reversible logic gate for quantum computing [113]. The T and H gate hereby act on one qubit by a change of its quantum states. The T-gate introduces a phase shift of $\pi/4$ between the two base states of the qubit. This can be done in optics for path entangle qubits by increasing the length of the optical path for one of the two photon paths that builds up the qubit. The H gate converts the two pure states of the qubit into intermixed states. In optics this is realized by the use of a 50:50 BS between both optical paths. For a more detailed description of the concepts behind quantum gate logic, the reader is advised to resort to more specialized literature [114–116].

The realization of a CNOT gate is more complex. A concept for the realization of a photonic coincidence controlled CNOT proposed by [117] is shown in fig. 2.2. The CNOT consists of three BS with a 1:3 and two BS with a 1:2 splitting ratio. A control qubit (c_0, c_1) is used to toggle a NOT operation on a target qubit (t_0, t_1). c_0 and c_1 are the modes of the two photons that constitute the control qubit and t_0 and t_1 are the modes of the photons of the target qubit. In addition, two unoccupied ancillary modes v_c and v_t are necessary. The gate causes a sign shift as the target modes are split and remixed in the interferometer formed by the BS structure. The operation is conditional on the presence of a photon in the c_1 mode and on a coincidence measured between the control and the target qubit. The probability of such a coincidence is one-ninth. Eight times out of nine, either the target, the control, or both do not contain photons [117]. The limited probability requires a post-selection of correct results and cannot directly be used for scalable integration. However, it can be upgraded to a heralded gate, which can be used for full integration [118]. Nevertheless, the described CNOT can be used as a testbed and was demonstrated in free space [17] [18] and integrated on-chip using WGs and directional BS [26] [30]. This shows that the full integration of quantum logic is possible using single-mode WGs and BS.

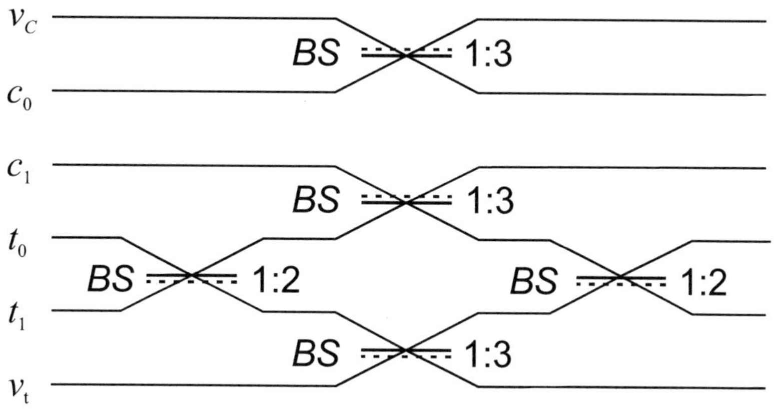

Figure 2.2: Coincidence-based photonic CNOT gate as proposed by [117]. The dashed lines at the beam splitters indicate where a phase shift of π happens upon reflection. The numbers indicate the necessary splitting ratio. c_0 and c_1 are the modes of the control qubit. t_0 and t_1 are the modes of the target qubit. v_c and v_t are unoccupied ancillary modes.

2.1.3 Waveguides

Photons in integrated circuits are generally guided by the use of WG cavities. For optical frequencies, the WG typically consists of a dielectric core with a higher refractive index, which is immersed in a dielectric with a lower refractive index. Both dielectric materials must be transparent to avoid the absorption of the guided photon. Due to the difference in refractive indices, the photon is confined and travels along the dielectric core. The strength of the confinement is dependent on the difference in refractive indexes of the used materials; the stronger the difference, the stronger the confinement. The optical modes that can propagate in the WG are dependent on the frequency f of the photon, the refractive indices n of the materials and the dimensions of the WG and can be calculated using Maxwell's equations. For a round WG typically used for optical fibers (fig. 2.3a), the E-field in the WG can be calculated by solving the eigenvalue problem

$$\nabla \times \nabla \times E = n(r)^2 \cdot k^2 E, \tag{2.1}$$

where $k = 2\pi f / c$ is the wavenumber and $n(r)$ the refractive index profile of the WG [119]. When the cross-section of the WG is reduced, the number of modes that can fit spatially and can propagate is also reduced. For a QPIC, the WG dimensions need to be chosen to support only the fundamental transverse electric (TE) or transverse magnetic (TM) mode to allow for a single-mode operation only. In a multimode WG, the excitation of multiple WG modes would lead to undefined input states, which in turn lead to unpredictable results for operations using the shown photonic gates. The corresponding mode profile is shown in fig. 2.4a. For integrated WGs, rectangular geometries are easier to fabricate and, therefore, preferred (fig. 2.3b-e). The WG is hereby fabricated using one lithography and etching step.

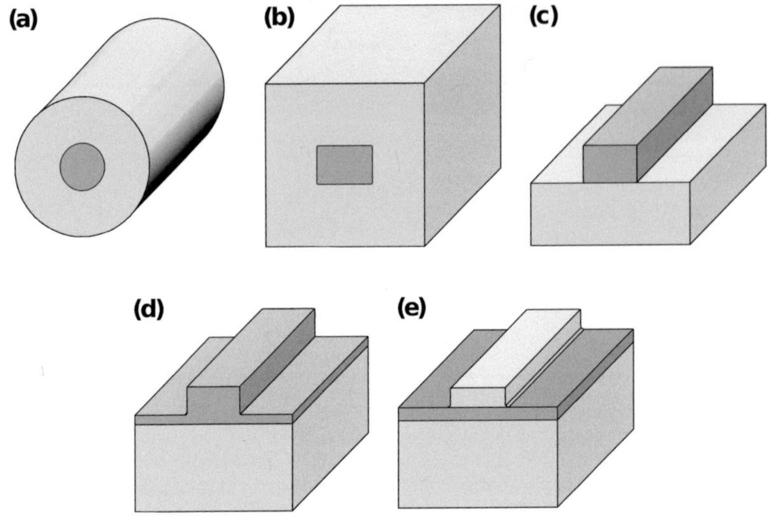

Figure 2.3: Overview of possible WG designs. The core (dark gray) has a higher refractive index than the cladding (light gray) and consequently confines the photon. a) Fiber WG. b) Channel WG. c) Ridge WG. The confinement of the photon towards the top is a result of the difference in refractive indices to the ambient medium. d) Rib WG, consists of a ridge WG with an additional slab. e) Strip-loaded WG. Reprinted by permission from U. Rengstl: Springer 2017 [16].

In the case of a channel WG, the etched ridge is overgrown in an additional deposition process. Channel WGs offer a mode profile better suited for detector integration, since a significant amount of the evanescent field is seen outside of the core on top of the WG (fig. 2.4b). However, integrated laser excited single-photon sources are best accessible in a free-standing WG structure. In a free standing structure, the top confinement is typically reached by the index contrast between the WG core and the vacuum. Since the index contrast to vacuum is higher than to a cladding dielectric, the mode propagates deeper inside the substrate. In the GaAs material system, the index contrast of the lower cladding and the WG core is obtained by the use of AlGaAs. The highest field density on top of a free standing WG is achieved by the use of a ridge WG (fig. 2.4c).

This, however, leads to an open AlGaAs cladding layer. This is problematic since AlGaAs is sensitive to degradation by oxidation [110]. Consequently, the preferable WG configuration is a rib WG structure with a remaining GaAs slab that protects the underlying AlGaAs layer. Another possible WG geometry is a photonic crystal WG. The waveguide is thus made of a continuous layer made from a single material into which holes are etched on order to lower the effective refractive index of the "cladding". The top and bottom cladding is done by the utilization of the index contrast to vacuum so the entire structure is basically a membrane with holes at the side. This offers an almost perfect coupling of QD emission to the WG [57] and removes losses in the cladding dielectric and, consequently, allows the creation of low loss WGs [120]. However, its membrane-like geometry makes large structures very fragile and difficult to handle [121]. This section was written with the help of [16]. Detailed investigations of rib WGs on GaAs can be found in [110].

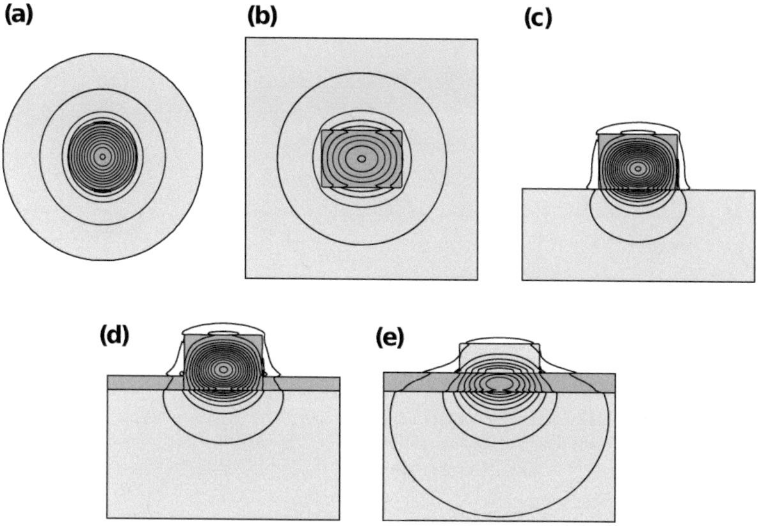

Figure 2.4: Profiles of the fundamental TE modes for the WG geometries shown in fig. 2.3 for a GaAs(dark gray)/$Al_{0.42}Ga_{0.58}As$(light gray) material system. The lines illustrates the E-field as $|E|^2$ in a linear scale. Reprinted by permission from U. Rengstl: Springer 2017 [16].

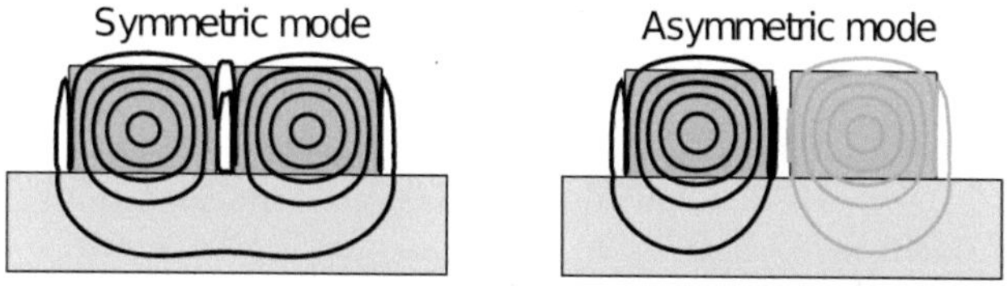

Figure 2.5: TE mode distribution for two ridge WGs forming a BS, for a symmetric and an asymmetric mode in the WG. Reprinted by permission from U. Rengstl: Springer 2017 [16].

2.1.4 Beam Splitter

A beam splitter can be created by the use of two on-chip WGs. A so-called directional coupler is created by bringing the WGs so close together that the evanescent fields of the WG modes overlap (fig. 2.5). Hereby, the mode couples between both WGs and the coupling ratio can be controlled by the interaction length of the coupler L and the distance (gap) between the two WGs. The propagating mode in the coupler can be described in the framework of the super-mode theory [110] by a super-mode which consists of a symmetric and an asymmetric mode of the propagating TE and TM mode, respectively. Since the symmetric and asymmetric mode have a different mode profile, they each "see" a different effective refractive index and the propagation constants of both modes vary, which leads to a beating between the two modes and an observable energy transfer between two WGs (fig. 2.6). The length of the coupler necessary to reach the wanted power transmission T from one to the other WG can be calculated as:

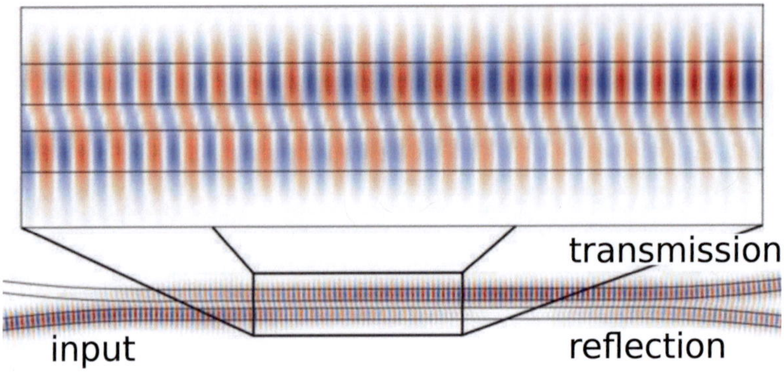

Figure 2.6: Simulated mode beating of the TE mode between the two WG arms of a 50:50 directional coupler. Negative values of the TE mode are depicted in blue and positive values in red. Under close inspection, a phase shift of $\pi/2$ introduced by the coupler is visible. The inset is a magnification of the coupler region. Adapted by permission from U. Rengstl: Springer 2017 [16].

$$T = \sin^2\left(\frac{L}{2}(k_{z_{\text{sym}}} - k_{z_{\text{asym}}})\right). \tag{2.2}$$

$k_{z_{\text{sym}}}$ is the propagation constant of the symmetric mode and $k_{z_{\text{asym}}}$ is the propagation constant of the asymmetric mode, respectively. Both propagation constants are dependent on the effective refractive index, which is itself dependent on the distance of both WG arms. The rest of the power gets reflected (R) at the beam splitter and stays in the originally excited WG. The reflection can consequently be calculated as $R = 1 - T$. To reach a splitting ratio of 50:50, the length can be calculated as:

$$L_{3db} = \frac{1}{2}\frac{\pi}{|k_{z_{\text{sym}}} - k_{z_{\text{asym}}}|}. \tag{2.3}$$

$k_{z_{\text{sym}}}$ and $k_{z_{\text{asym}}}$ are necessary for the calculation and can be extracted out of frequency domain simulations. A detailed description of this type of simulations along with simulation results can be found in [110]. A potential problem of using directional couplers for high yield fabrication is the gap width. To have an overlap of the evanescence field, the gap width is typically in the order of 100 nm. Even small changes in the width lead to a significant change of the coupling ratio, which makes low fabrication tolerances and a tight control of the lithography and etching steps necessary.

An alternative to directional couplers are multi-mode interference couplers. They operate by transforming the single-mode existing in the WG to multiple modes by a local increase of the WGs width. The modes interfere with each other and the desired splitting ratio is obtained by a placement of an output single-mode WGs at a position with the aspired energy density

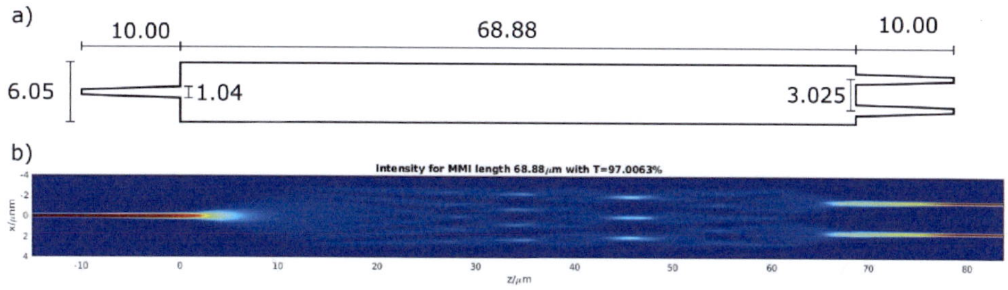

Figure 2.7: Multi-mode interference (MMI) BS with one input and two output WGs with a 50:50 splitting ratio. a) Dimensions of the MMI in µm. b) E-field in the MMI, several mode maximums and minimums are visible. Reprinted (adapted) with permission from [70]. Copyright 2018 American Chemical Society.

(fig. 2.7). Due to a larger critical width, the splitting ratio of the MMI is more tolerant to geometrical variations caused by the fabrication processes. However, fabrication derivations cause additional losses and back reflections in the MMI body. In a direction coupler on the other hand, losses are only determined by the attenuation of the used WGs and since the WG widths remain stable, back reflections are avoided. Furthermore, the use of variable phase shifters allows a correction of the splitting ratio of a directional coupler by the using a Mach-Zehnder interferometer design [110]. Hence, directional couplers were used for full integration during this thesis.

2.2 Semiconductor Quantum Dots

Quantum dots (QDs) are structures that tightly confine charge carriers in all three spatial dimensions. The confinement ideally forms a zero dimensional potential trap with an ideal two-level system that allows charge carriers to exist only in one discrete energy level. Due to the Pauli exclusion principle, the amount of electrons at this energy level is limited to two with opposing spins. In analogy to the energy shell of an atom this fundamental energy level is called the s-shell. When an electron-hole pair (exciton) that is excited in such a system recombines radiatively, a photon is emitted with an energy that corresponds to this fundamental energy level. Furthermore, due to the single existing energy state, only one photon is emitted at a time. In the GaAs material system, the large band-gap difference between InAs ($E_G(2\,\mathrm{K}) = 0.422\,\mathrm{eV}$) and GaAs ($E_G(2\,\mathrm{K}) = 1.522\,\mathrm{eV}$) is used to create the necessary three dimensional confinement. The Stranski-Krastanov growth [122] of InAs on a GaAs(100) initially results in a two dimensional wetting layer However, due to the lattice-mismatch between GaAs and Ingas, the InAs continues to grow in a pyramidal, three-dimensional shape. During the capping with GaAs, the capping layer intermixes with the QD and a InGaAs/GaAs QD is formed [60]. The resulting QD resembles a dome shape which leads to a parabolic carrier confinement that can be approximated by a quantum well in the z-direction and a harmonic oscillator in the plane of the substrate (xy-plane). This generates an only quasi-zero dimensional confinement and, in dependence on the size of the dot, higher energy shells (p, d ...) can exist with a number of electrons corresponding to the number of

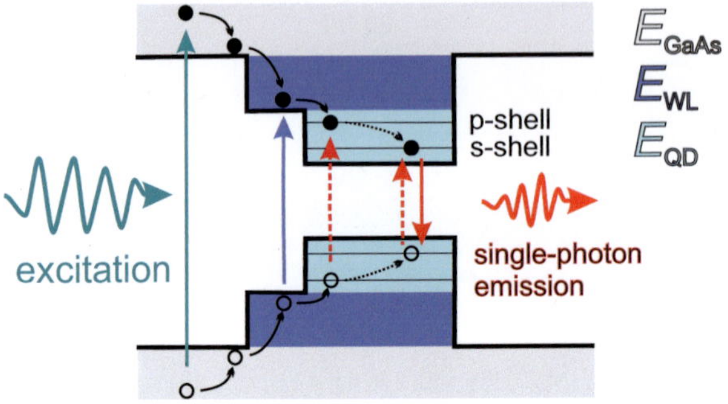

Figure 2.8: QD band structure (not to scale) along its z-axis. Upwards arrows symbolize off-resonant (solid arrow) and quasi resonant (p-shell) and resonant (s-shell) (dashed arrow) excitation processes. The downward arrow symbolizes the radiative emission processes. Black arrows symbolize relaxation processes for electrons (filled dot) and holes (open dot). The figure was drawn after [16].

shells. The n-th shell can hold up to $2 \cdot n$ electrons [123]. The spatial dimensions are in the range of the De-Broglie wavelength of the electron:

$$\lambda_{\text{De-Broglie}} = \frac{h}{\sqrt{2 \cdot m_e^* \cdot E_{\text{therm}}}}, \tag{2.4}$$

where m_e^* is the effective electron mass, which is material dependent, h the Planck constant and $E_{\text{therm}} = k_B T$ the thermal energy of the electron where k_B is the Bolzman constant and T is the temperature. Using the InGaAs/GaAs material system, QD emission wavelengths in the range of 850 to 1400 nm, can be achieved [60].

2.2.1 Optical Excitation of a QD

In the following section, a QD with a size that supports the fundamental s- and p-shell is considered. The corresponding band structure is sketched along its z-axis in fig. 2.8. The GaAs host material is colorized in gray. Since in the indium wetting layer mixes with GaAs, the bandgap is slightly reduced (blue). The significantly decreased bandgap inside InGaAs QD is shown in cyan. The QD can have several excited states. Electron and holes that are generated by optical pumping can diffuse and get trapped into the QD and relax to its lowest energetic state, the s-shell. When confined, they Coulomb interact and form a quasi-particle, called exciton (X). This exciton can recombine radiatively by the emission of a photon. The typical lifetime of an exciton is in the sub-nanosecond range. Recombination processes are only allowed with electrons and holes in the same shell. Since the s-shell can hold two carriers with opposing spins, two electron-hole pairs can exist, constituting a so-called biexciton (XX); the total spin of the XX is zero. When a single carrier exist in the s-shell in combination with an X, they constitute a so-called trion, with either an excess electron X^- or an excess hole X^+. XX and trion states are detuned in energy in comparison to the X state due to

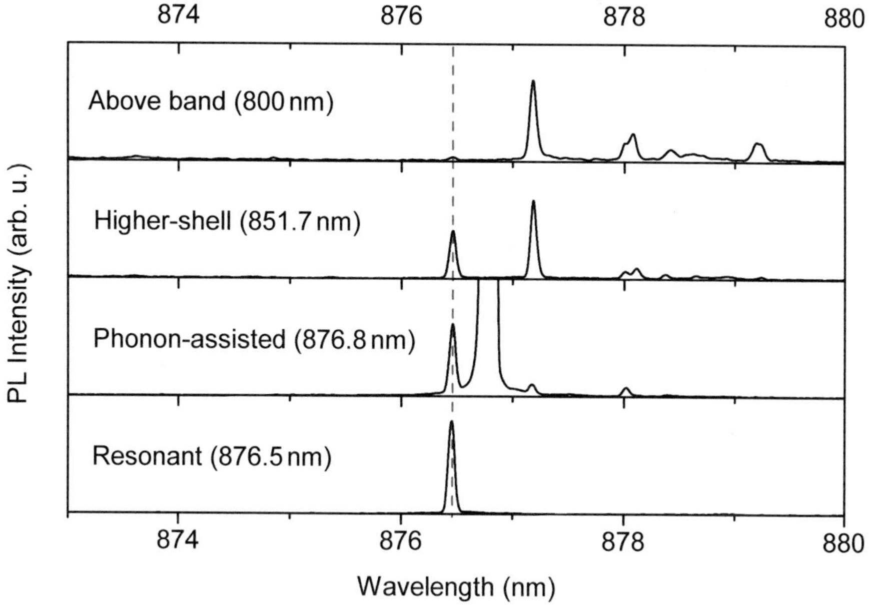

Figure 2.9: QD emission into a WG for different excitation mechanisms. The scaling for all graphs is identical, the excitation power is different. The graph was taken from [124].

Coulomb interactions. Several excitation mechanisms can be used to create an X (fig. 2.8), in dependence on the excitation energy. [16] Off-resonant excitation schemes use excitation energies significantly higher than the bandgap energy of the QD and can excite carriers in either the GaAs or the wetting layer. They relax into the QD s-shell by releasing energy due to phonon interactions. This kind of excitation is further regarded during this thesis as above-band excitation. Above-band excitation is easy to implement since the used pump does not need a well-defined energy. In general, light from an above-band pump laser can be filtered from the QD emission by the use of spectral filtering as the energy of the emitted photon is significantly smaller than the energy of the pump photons.

Disadvantages for above-band excitation are a larger jitter for the emission connected to the relaxation processes necessary until the X reaches the s-shell. The excitation probability of multi-exciton states during above-band excitation is increased, which limits the coherence of emitted photons. Both effects limit the indistinguishability of excited photons [125]. In addition, above-band excitation does not allow the deterministic excitation of a single-QD. The high photon energy of an above-band excitation laser in comparison to the QD energy level also excites QDs nearby, which introduces unwanted photons into an experiment (fig. 2.9). To reduce the number of necessary relaxation processes, the laser energy can be tuned to the energy of the p-shell in quasi-resonant excitation, thus limiting the number of intermediate relaxation steps to one. This excitation scheme results in a significantly improved coherence of the emitted photons [126]. Best results, however, are obtained by a resonant excitation of the s-shell of the QD. The photon energy of the excitation laser hereby corresponds to the energy level of the QD. This leads to spectrally clean emission (fig. 2.9) and allows a high

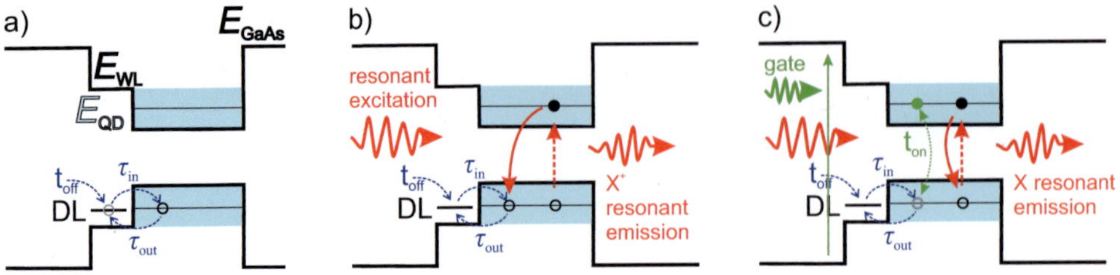

Figure 2.10: Optical gating of a resonantly excited QD after [128]. a) With the resonant excitation laser switched off, a hole from a structural deep level (DL) in vicinity of the QD can tunnel in and out at characteristic time constants τ_{in} and τ_{out}. A filling of the DL at a rate of $1/t_{\text{off}}$ leads to a residual hole in the QD. b) The QD is detuned and only the trion state X^+ can be optically exited and emit photons. The optical emission of the X is blocked. c) The addition of an above-band gate laser generates electrons that recombine with the residual hole at a time constant t_{on}. If $t_{\text{on}} < t_{\text{off}}$, the QD remains in a neutral state and emission of photons with energies of the neutral X is visible.

degree of indistinguishability of the emitted photons [127] [16]. Furthermore, it makes the selective excitation of a single-QD possible. For nearby, slightly off-resonant QDs either the energy is to small to excite an X or the excess energy in comparison to the band gap is to small to go to another process. The efficient resonant excitation is challenging and requires a tight control of the spectrum and the polarization of the excitation laser. In addition, a gate laser has to be used in most cases in order to gain stable resonant emission out of a QD using coherent excitation (fig. 2.10).

A defect in close vicinity of the QD can cause a structural deep level (DL). A hole tunneling from the DL to the QD quenches the QD resonant emission by a Coulomb blockade [129]. When the resonant laser is switched off, a hole from the DL can tunnel in and out at characteristic time constants τ_{in} and τ_{out}. The DL is filled at a rate of $1/t_{\text{off}}$ which leads to a residual hole in the QD. When the QD is resonantly excited, the residual positive charge of the hole causes a Coulomb blockade that detunes the QD and blocks the resonant excitation of the X. Only a trion state can be excited to emit photons. In order to solve this, an optical gate laser at an above-band gap energy with low intensity can be used to photogenerate an electron in the surrounding GaAs substrate. This additional electron recombines with the residual hole at a time constant t_{on}, neutralizing the QD. When $t_{\text{on}} < t_{\text{off}}$, the QD is stabilized in its neutral state and the wanted resonant photon emission generated by a neutral X-state is visible [128]. The necessary excitation scheme and further details into resonant excitation are discussed in detail in [16] [124].

2.2.2 2nd-Order Correlation

Typically, the emission of a photon source is characterized by a measurement of its 2nd-order correlation function $g^{(2)}(t)$ [90]. $g^{(2)}(t)$ gives the likelihood of a two-photon correlation of a given light source. It is measured by a time correlation of detection events of two single-photon detectors and can be done without knowledge of the detector efficiencies. A typical $g^{(2)}(t)$ measurement is performed in a Hanbury-Brown and Twiss setup (HBT) [107]. The

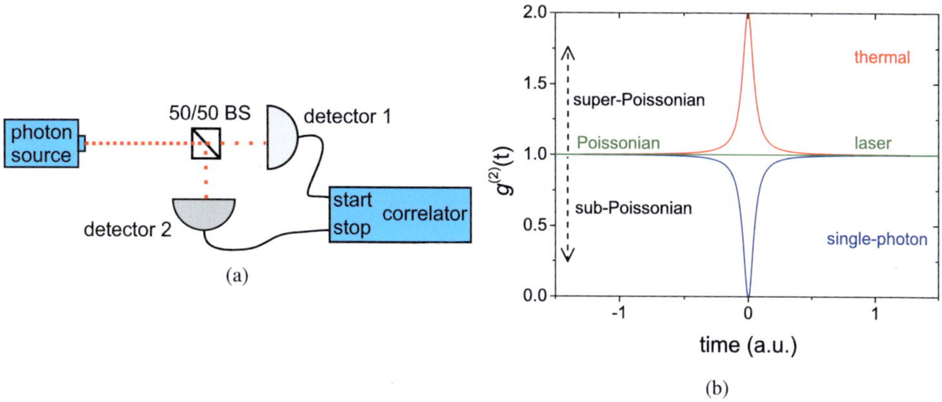

Figure 2.11: (a) Schematic of a Hanbury-Brown Twiss setup [107]. (b) Two-photon correlation for different photon sources. At $t = 0$ thermal emission shows bunching while single-photon emission shows antibunching.

setup is schematically displayed in fig. 2.11a. It consists of two single-photon detectors in combination with a BS. The highest efficiency during the measurement is achieved by the use of detectors with an identical detection efficiency and a 50:50 splitting ratio of the beam splitter. For the measurement, the first detector is used as a time reference and a timing histogram for detector counts of the second detector is recorded. The histogram is normalized to unity for the average intensity, which leads to a histogram in a form shown in fig. 2.11b. If one photon is emitted from a thermal light source, the probability for the emission of a further photon rises. Hence, there is a higher probability of the second detector also registering photons if the first detector detects a photon. As a result a peak can be seen in the histogram This effect is called bunching and the photon emission statistics is super-Poissonian. In a laser, the emission of photons happens independent from each other which causes its Poissonian emission characteristic and a $g^{(2)}(t) = 1$ over the full time. An ideal single-photon source, however, emits only one photon at a time and if a photon is detected by the first detector, a photon can never be detected on the second detector at the same time and correspondingly a dip to zero can be seen, $g^{(2)}(0) = 0$. Its characteristic is correspondingly sub-Poissonian. For a real single-photon source, some degree of multi-photon emission is always present. Consequently, the $g^{(2)}(0)$-value is used as a quality parameter to describe how close the measured single-photon source resembles an ideal single-photon source ($g^{(2)}(0) = 0$) [111]. If anti-bunching is present and the $g^{(2)}(0) < 0.5$, the investigated source is regarded to be a single-photon source. A fully on-chip HBT setup is used in this thesis as a test bench to demonstrate all elements required for QPIC on a single-chip, since it consists of all basic elements required for LOQC and to allow the characterization of single-photon operation.

Figure 2.12: Scanning electron microscopy (SEM) image of a meander SNSPD, the superconductor is seen in black.

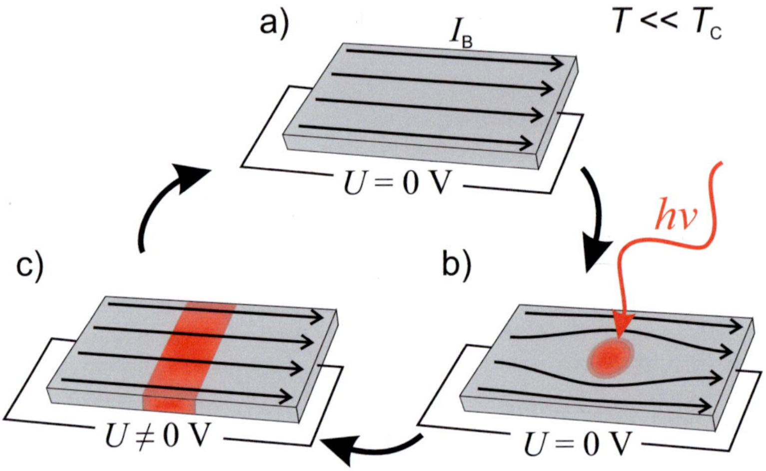

Figure 2.13: Schematics of the SNSPD response mechanism to the absorption of a photon. a) Initial state: $T << T_C$, $I \leq< I_C$; b) Photon absorption in the nanowire; c) The formed resistive barrier causes a voltage potential along the nanowire.

2.3 Superconducting Nanowire Single-Photon Detector (SNSPD)

A typical SNSPD consists of an ultra-thin superconducting nanowire with a thickness d = 4 to 6 nm and a width $w \leq 100$ nm. To increase the detection area, the nanowire is typically folded in a meander shape (fig. 2.12). The detection of photons in a SNSPD is based on the quenching mechanism for superconductors. Superconductivity in a superconductor can only exist if three critical conditions are fulfilled: the temperature T, current I and magnetic field H must be below the materials critical temperature T_C, critical current I_C and critical magnetic field H_C. Once these conditions are fulfilled, electrons bound together in pairs, so-called Cooper pairs, that can move through the phonon lattice without resistance. For detector operation (fig. 2.13), the nanowire must be cooled to temperatures well below its critical temperature $T << T_C$. A bias current I_B close to the critical current I_C is applied (a), the nanowire is superconducting and no initial voltage can be measured along the wire. When a photon with sufficient energy is absorbed in the nanowire (b), the nanowire quenches locally, gets normal conducting (c) and a normal conducting zone grows due to joule heating. Since the quenched zone is normal conducting and a current is applied, a voltage can be measured along the wire. As the detector has a high resistance in the quenched state, the current is redistributed over an external shunt and the nanowire can recover to its initial state, ready to detect another photon [130]. Fast quenching and recovering times cause detection events to be visible as a voltage pulses with a pulse length in the order of nanoseconds. The count rate is limited by the dead time, which is defined as the time the detector needs to recover after a detection event.

2.3.1 Detection Efficiency

The probability of detecting a photon for a SNSPD integrated into an optical setup is given by the system detection efficiency (SDE):

$$\mathrm{SDE} = \mathrm{OCE} \cdot \mathrm{ABS} \cdot \mathrm{IDE}. \tag{2.5}$$

OCE is the optical coupling efficiency, which is the probability of the photon reaching the detector through a given optical path; ABS is the probability of a photon to be absorbed by the SNSPD after reaching the detector; IDE is the intrinsic detection efficiency, which gives the probability of a photon that is absorbed in the detector to trigger a voltage pulse. Since the OCE is a parameter that describes the full optical system, the detection efficiency (DE) of the detector is defined as:

$$\mathrm{DE} = \mathrm{ABS} \cdot \mathrm{IDE}. \tag{2.6}$$

Experimentally, the DE is characterized as:

$$\mathrm{DE} = \frac{N_{\mathrm{counts}} - N_{\mathrm{darkcounts}}}{N_{\mathrm{photons}}}, \tag{2.7}$$

where N_{counts} is the amount of detector counts and $N_{\mathrm{darkcounts}}$ the amount of detector dark counts. N_{photons} is the number of photons on the active area of the detector. For a waveguide SNSPD (WG SNSPD) illuminated through the WG, N_{photons} is the amount of photons in the WG at the position of the detector.

The ABS can be derived from the Beer-Lambert law as:

$$\mathrm{ABS} = 1 - e^{-\frac{4\pi\kappa l}{\lambda}}. \tag{2.8}$$

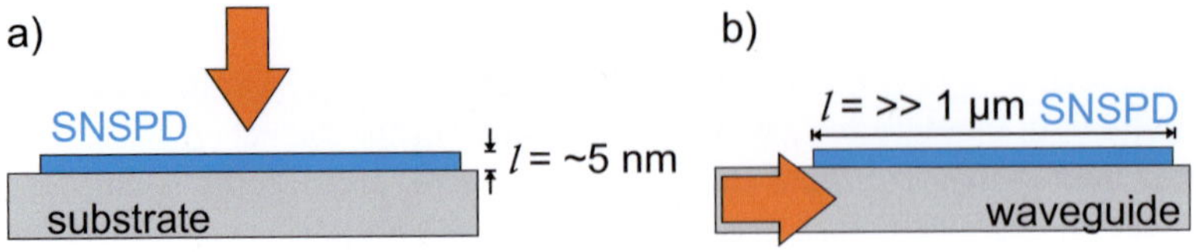

Figure 2.14: Absorption length for top illumination a) and WG illumination b) of the detector.

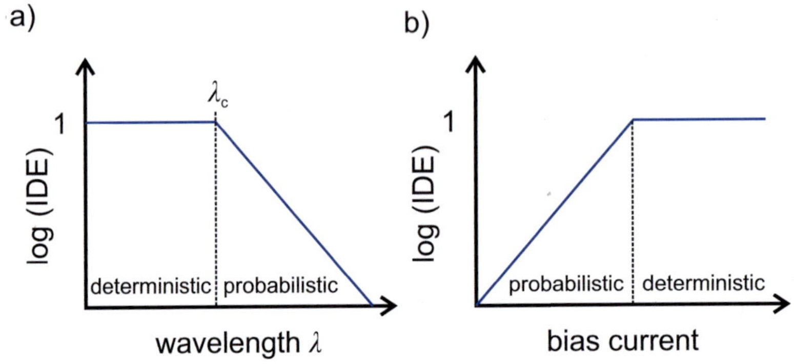

Figure 2.15: a) Simplified wavelength dependence of SNSPD detection regimes. The cutoff wavelength λ_C is sketched. b) Simplified bias current dependence of SNSPD detection regimes.

κ is the attenuation coefficient, which is the complex part of the effective refractive index (mode index), l is the absorption length and λ the wavelength. The l of a SNSPD optically excited from the top is limited to the thickness of the film (fig. 2.14b). Hence, the ABS for a $\approx$5 nm thick NbN SNSPD is limited to $\approx$30 % [131]. For a WG integrated SNSPD, l is dependent on the length of the SNSPD on the WG. Typically, an l of several micrometers can be realized (fig. 2.14b). The ABS can therefore be pushed towards unity for a sufficient detector length on an ideal WG with zero internal absorption. While the effective κ on a WG is almost three orders of magnitude smaller than the κ of the detector itself [132], the effect is outweigh by an l that is easily four orders of magnitudes longer. A high ABS with small l can be obtained by the use of an optical cavity around the detector [37], which causes a photon to hit the SNSPD several times until it is either absorbed or lost from the cavity.

The IDE is strongly dependent on the detection regime of the SNSPD. Depending on the photon energy, the SNSPD operates in one of two detection regimes (fig. 2.15). The first regime is the deterministic regime: the absorbed photon has sufficient energy to trigger a detection event by itself and triggers a detection event deterministically. Since each photon triggers a detection event, the IDE in this regime is close to unity. The second regime is the probabilistic regime: the absorbed photon lacks sufficient energy to trigger a detection event and must be assisted by a vortex. Since the probability of a photon to trigger a detection event is dependent on the presence of a vortex, the IDE is proportional to the probability of a vortex to enter the nanowire. The detection regime in dependence on the wavelength of the photon is sketched in fig. 2.15a. The λ where the detection switches from the deterministic to the probabilistic regime is defined as the cutoff wavelength λ_C. Since the roll-off that is experimentally seen is slow and there is usually no sharp transition between the detection regimes, λ_C is experimentally determined as the λ where the DE drops to 50 % of the DE in the deterministic detection region.

2.3.2 Detection Mechanism

The dependence of the cutoff wavelength on the detector parameters and on external parameters can be qualitatively described in the framework of the diffusion-based hot spot model [133]. In this model, the number of current carrying Cooper pairs is considered in a section of the wire at a length corresponding to the Ginzburg-Landau coherence length ξ_{GL}. The absorbed photon excites an electron to an energy level significantly larger than the superconducting gap Δ in the nanowire. The electron thermalizes in a cascading process due to electron-electron and electron-phonon interactions under generation of a large number of quasiparticles (QPs) in a thermalization time τ_{th}. The redistribution of energy caused by the QP multiplication process breaks up Cooper pairs. Since the supercurrent remains constant but is carried by a lower number of Cooper pairs, the remaining Cooper pairs have to speed up. If enough Cooper pairs are broken, the remaining pairs exceed a critical velocity, they break up and a normal conducing zone occurs in the nanowire [134]. The minimum energy E_{min} and ,correspondingly the longest λ_C to deterministically detect single photons can be described in dependence of the applied bias current I_C as:

$$E_{min} = \frac{hc}{\lambda_C} = \gamma \cdot \left(1 - \frac{I_B}{I_{dep}}\right). \tag{2.9}$$

The depairing current I_{dep} is the current necessary to split all Cooper pairs, h is the Planck constant and c the speed of light in vacuum. γ is a device- and material-dependent proportionality factor and can be expressed as [133]:

$$\gamma = \left(\frac{N_0 \Delta^2 w d}{\varsigma}\right) \sqrt{\pi D \tau_{th}}, \tag{2.10}$$

where N_0 is the electron density of states, D is the electron diffusion coefficient and ς is the efficiency of the QP multiplication process. One can see that within this model, the ratio I_B/I_{dep} is key to a reach a detector that is sensitive towards long λ_C. This model is in good agreement with several experimental observations: A linear dependence of the cutoff wavelength on the thickness d and width w is observed [135, 136]. It also gives a reasonable description of the observed dependence on the material-dependent parameters N_0, D and Δ [137]. When taking into account the temperature dependence of Δ and D, the temperature dependence described in the model is qualitatively correct for $T/T_C \lesssim 0.5$ [138]. However, it predicts a shift of λ_C to lower photon energies for $T/T_C > 0.5$ which is not experimentally observed. For an operation temperature of the $T \rightarrow T_C$, the photon energy dissipates before a normal conducting zone

can be created, which is causes the λ_C to decrease for operation temperatures close to T_C. As a consequence, the operation temperature of a SNSPD should be lower than half of its critical temperature $T < 0.5T_C$ [139].
The diffusion-based hot spot model can only describe the detection of photons in the deterministic detection regime. In the probabilistic regime, the exponential energy dependence of the IDE hints to a photon detection that is occasionally assisted by thermal or quantum mechanical fluctuations. A natural candidate for these fluctuations are magnetic vortexes that assist the local destruction of superconductivity after the absorption of a photon. Two mechanisms for vortex entry are generally considered: The first mechanism is a photon-triggered unbinding of a vortex-antivortex pair in the nanowire and the second mechanism is vortex entry into the nanowire from the side, which is made possible by a lowered potential barrier (vortex entry barrier) due to photon absorption. Once inside the nanowire, the Lorentz force caused by the applied bias current acts on the vortex and causes the vortex to cross the nanowire. During the vortex movement, energy is dissipated and a detection event is triggered. The potential barrier for vortex-antivortex unbinding is significantly higher than for a single-vortex crossing [140]. An exponential dependence of the vortex entry probability on the height of the vortex entry barrier [141] explains the observed exponential dependence in the probabilistic regime. An overview of different detection mechanisms of SNSPD and a comparison with experimental observations can be found in [134]. A recent comparison of the diffusion-based vortex-entry model and the normal-core vortex model is given in [139].

2.3.3 Dark Counts

Dark counts in a SNSPD are all counts that are triggered without the presence of single photons. Dark counts caused by electronic noise can be omitted by the use of low noise electronics and proper shielding. Intrinsic dark counts are understood to be caused by single-vortices crossing the nanowire [142]. As for the vortex assisted detection of single photons, thermally activated vortices with sufficient energies may overcome the vortex entry barrier without the presence of a photon. Once inside the nanowire, the Lorentz force, caused by the applied bias current, acts on the vortex causing the vortex to cross the nanowire. The movement of the vortex dissipates energy in the wire and a detection event is triggered. Since these vortices are thermally activated, the corresponding intrinsic dark count rate is exponentially dependent on the barrier height [140]. Since the height of the barrier is proportional to the energy gap in the nanowire, intrinsic dark counts have an exponential dependence on the applied bias current. Structural defects in the nanowire can lower the vortex entry barrier and increase the observed number of dark counts. This leads to an increased number of dark counts in detectors with sharp turns due to current crowding [143] [144]. Intrinsic dark counts can be decreased significantly by an improved thermal coupling [145] or a lower operation temperature [146] of the detector. For an application, the quoted dark count rate (DCR) often includes all counts not triggered by the investigated photon source. Besides intrinsic dark counts and electronic noise, this also includes photon counts from the thermal background and from stray light. Within the framework of this thesis, the DCR includes intrinsic dark counts, electronic noise and thermal counts, while stray light is mentioned separately. When stray light is included the

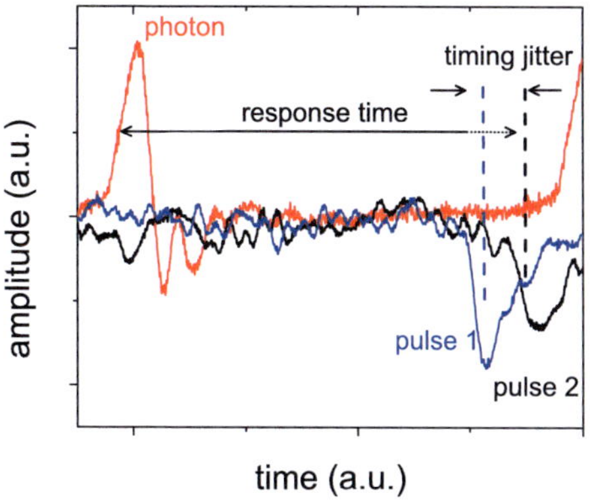

Figure 2.16: Detector response time (latency) and timing jitter.

term noise counts is chosen. A detailed investigation of noise counts for a QPIC using laser excited QD photon sources can be found in section 7.3 of this thesis.

2.3.4 Timing Characteristics

For the timing characteristics of the detector, two parameters are of interest: the dead time of the detector and the timing jitter. The dead time is the time that has to elapse after a detection event until the detector has recovered and is ready to detect another photon. It defines the maximal count rate of the detector. The dead time is dependent on a time constant τ caused by the kinetic inductance L_{kin} and the external shunt resistor R_P of the nanowire $\tau = L_{kin}/R_P$ [147] and the thermal relaxation time. The thermal relaxation time is typically in the order of tens to hundreds of picoseconds and is significantly smaller than τ, which is typically in the order of nanoseconds [148]. L_{kin} is mainly determined by the length of the nanowire and the used material. If τ is too small, the nanowire cannot fully recover, which for a standard SNSPD design causes the detector to stay in a stable resistive domain and the detector latches [149]. To avoid latching in very short nanowires, additional on-chip inductors can be added which slows down the return of the bias current.

The timing jitter describes the timing uncertainty of the response time (latency) of a SNSPD for the detection of a photon (fig. 2.16). It gives the time resolution of the detector. The intrinsic timing jitter is dependent on the geometry and detector material. The underlying physical mechanisms are not fully understood and are the focus of many researchers in the recent years [150–153]. For an optimized differential readout scheme in combination with very short nanowires, jitter values as low as 3 ps have been demonstrated. In an application, the used RF-readout heavily influences the measured jitter. In this thesis, the measured jitter values are given as system jitter values, which gives the full instrumental response function (IRF) of the full setup including the optical excitation and readout. In a QPIC, the visibility of photon correlations is better for a lower jitter.

2.3.5 SNSPD Material

The performance of a SNSPD is dependent on the used superconductor. Materials for SNSPDs with a good ability to detect single-photons are extremely dirty superconductors with small electron-diffusion coefficients ($D_{NbN} \approx 0.5\,\mathrm{cm^2/s}$) and low critical temperatures. A small D does not allow the fast diffusion of electrons and causes a strong local suppression of Δ. Most of the photon energy needs to be distributed in the electron system, hence the heat capacity of the electron system for the material should be large in comparison to the heat capacity of the phonon system. A low T_C material has a small Δ and needs a lower energy to be quenched. The operation temperature for the detector to work efficiently should be $< 0.5T_C$. At temperatures close to T_C, the retrapping current I_r is close to I_C which causes the hotspot to shrink and no normal conducting region can form across the wire [139]. Typical materials used for SNSPDs are nitrides (NbN, TaN, NbTiN) and silicides (WSi, MoSi) [154]. Nitrides are crystalline materials that offer a higher critical temperature, a lower timing jitter and faster recovery times, while silicides are amorphous materials with a higher sensitivity for lower photon energies. NbN-based SNSPDs typically have a $T_\mathrm{C} \approx 10 \pm 2\,\mathrm{K}$, which enables operation using cooling in liquid helium at 4.2 K. As a crystalline material, the growth properties of NbN have a strongly influenced by the substrate material and the growth conditions and careful optimization of the growth process is required (chapter 3). WSi-SNSPDs have a $T_\mathrm{C} \approx 4\,\mathrm{K}$ and, consequently, require operation temperatures of <2 K to be efficient, which requires a complex and expensive cryostat. The amorphous film structure allows an easier fabrication of uniform films with a higher tolerance on the used substrate and growth conditions [155]. The lower efficiency but better timing characteristic of the NbN film can be explained by the shorter hotspot lifetime in the range of 22 ± 1 ps [156] and a smaller size of the hotspot in comparison to 630 ps in WSi [155]. In the scope of this thesis, NbN was chosen as detector material because of its better timing characteristics and higher T_C that allows detector operation at 4 K.

2.3.6 Geometry

The geometry of the detector is an important design parameter. An adaptation of the geometry allows the design of application-specific SNSPDs by the optimization of particular parameters. To get a large $I_\mathrm{B}/I_\mathrm{dep}$, the nanowire needs to have an uniform I_C over its full length. Hence, a variation in I_C limits the maximal usable I_B to the smallest local I_C. This requires the nanowire to be as uniform as possible. The cross-section of the nanowire gives the volume in which the superconductivity has to be suppressed for a successful detection event. A smaller cross-section increases λ_C. The width (w) is limited by the needed pulse height and the influence of fabrication defects. To have a high enough SNSPD pulse and thus to get a good noise separation, the I_C of the full detector should be $>10\,\mu\mathrm{A}$. The edge roughness of the nanowire after fabrication is typically in the range of 5 nm. A smaller w causes a larger relative variation of the local current density and the nanowire gets less uniform. The influence of edge defects get more significant which increases dark counts. Typically, a w used for NbN SNSPDs is $\approx$100 nm. The nanowire thickness d can be fabricated to be significantly more uniform, which makes small thicknesses in the range of 4 to 6 nm feasible. Smaller thicknesses decrease the T_C and decrease the absorption. At larger thicknesses, a strong decrease of λ_C is observed.

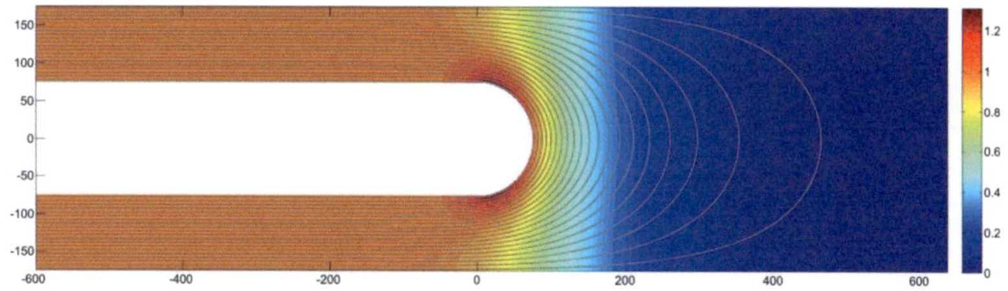

Figure 2.17: Simulated current density distribution in a typical bend of a meander SNSPD. The image was taken from [134].

Hence, 4 to 6 nm are the optimal thickness [135]. The length of the nanowire is chosen long (several hundred micrometer) if a large absorption area is needed and short (tens of micro meter) if a low dead time or a small timing jitter is required. To obtain a high ABS on a defined area, the nanowire is typically designed in a meander or more recently in a spiral shape [157]. Meander designs can be utilized if linear polarization sensitivity is required [135], while spiral designs are insensitive to linear polarization. Since spirals have either a turning point in the middle in the case of a double-spiral design [157] or a contact point in a single-spiral design [158], spiral designs suffer from an insensitve region in their middle. Strong bends in superconducting wires cause the presence of a high current density region on the inside of the bend [159]. This leads to a lower I_C in the bend and consequently lowers I_B/I_{dep} and, therefore, decreases λ_C. This effect is called the current-crowding effect [143].

2.3.7 Current-Crowding Effect

The current-crowding effect in superconducting strip conductors was first described in 1963 by Hagedorn and Hall [159], who observed that *"localized high current density regions near a right-angle bend can reduce substantially the critical supercurrent of such strip conductor"*. The effect was theoretically investigated for different planar geometries of bends by Clem and Berggren [143]. In asymmetric superconducting structures with a thickness d in the range of or smaller than the superconducting Ginzburg-Landau coherence length ξ_{GL} and a width w significantly smaller than the pearl length Λ, which is the case for superconducting nanowires, the measured critical current I_C is suppressed in respect to the depairing current I_{dep}. The suppression is stronger for sharper angles and smaller bend radii [157] [160]. The simulated current density contribution for a typical bend with a wire width w = 100 nm and a bend radius r of 75 nm is shown in fig. 2.17 [134]. The high current density in the inner bend reduces the energy barrier for vortex entry into the nanowire [142] and, consequently, the dark count rate of the detector [146]. The influence of the bend geometry on dark counts was shown in [144]. In accordance with the diffusion hotspot model [133], the cutoff wavelength λ_C for which photons can be detected deterministically is directly related to the ratio I_B/I_{dep}. Since the critical current of the detector is limited by its bends, the degree to which the superconducting energy gap can be suppressed in the nanowire is decreased. This directly reduces the maximal cutoff wavelength of the detector. To avoid current crowding, bends need to be designed with an optimal curvature. Since this requires larger bend radii, the filling factor is reduced to 1/3

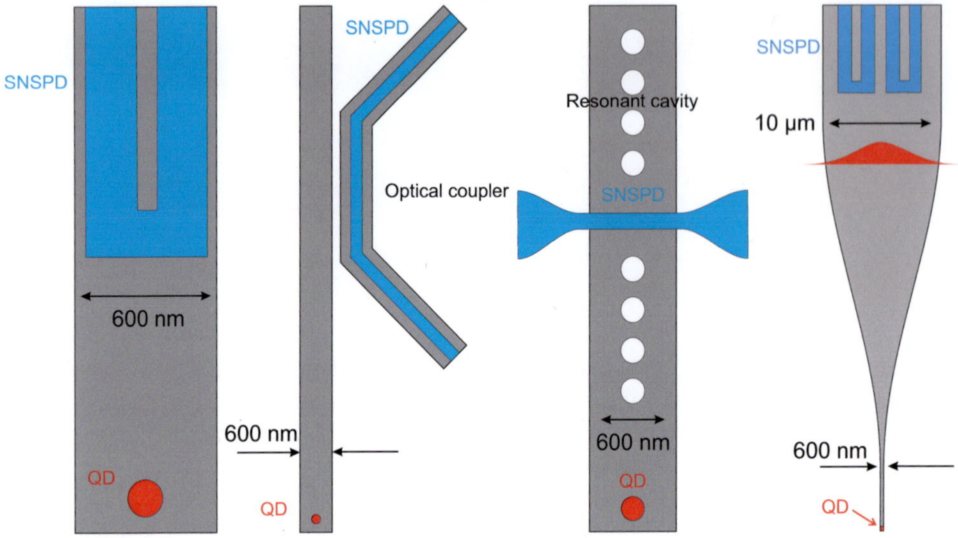

(a) U-shaped design (b) Coupler design (c) Resonant cavity design (d) Taper design

Figure 2.18: Different designs for a waveguide SNSPD. The dimensions in the graphics are not to scale.

for a meander detector with bends with an optimal curvature [143]. Another possibility to reduce current crowding is to use a spiral design in either double spiral [161] or single spiral configuration [158]. By using a single-spiral design instead of a meander design the I_C to I_{dep} ratio was increased from 0.40 to 0.55, with an increase of the corresponding λ_C from ≈600 nm to 900 nm [158].

2.3.8 Waveguide SNSPDs

The design of a SNSPD for quantum-photonic integrated circuits presents several challenges: it should enable a high DE for the used single-photon source (in this thesis for λ = 900 nm). The integration of the detector into a single-mode WG, with a width of ≈600 nm(for GaAs [110]), needs to be possible. To reach a high DE, the design should avoid current crowding. Several designs to embed a detector onto a small waveguide can be utilized. The layout used by most groups for waveguide detectors is an U-shaped design [162] [163] [35]. The design is schematically shown in fig. 2.18a. This design is simple and can be used without any modifications to the waveguide on-chip. The optical coupling of photons from the waveguide to the detector can be enhanced by increasing the length of the detector on the WG [162]. The disadvantage of this design is its 180° turn which limits the DE of the detector due to current crowing. In this configuration, the absorption for the WG-TE mode, which is the mode typically excited by an integrated self aligned QD [67], is significantly lower than for the TM mode [36]. Other designs rely on a modification of the waveguide-based photonic structure. The nanowire can be placed onto one arm of a directional coupler (fig. 2.18b) in a loop-type structure. In this design, sharp bends can be avoided and the coupler can be designed to have

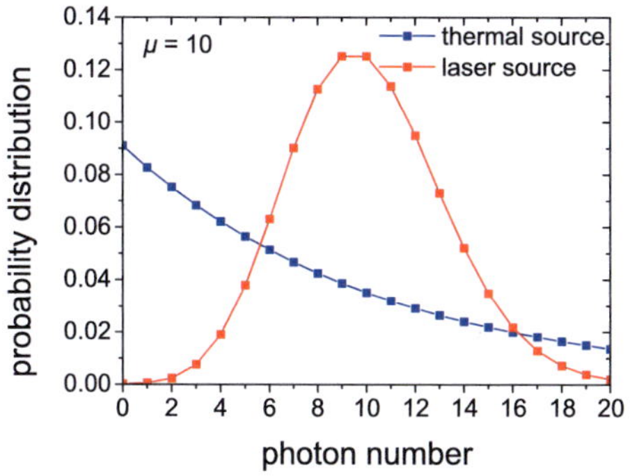

Figure 2.19: Photon statistics for a thermal (Bose-Einstein) and a laser (Poissonian) photon source for an average photon number $\mu = 10$. An ideal single-photon source will only have a photon number of one.

a defined coupling ratio, allowing this design to be used to monitor the single-photon flux in a WG at a defined loss. However, the coupling ratio is wavelength-specific and even small fabrication errors will shift the coupling ratio away from the design ratio. A third possible design is the embedding of the SNSPD into a phonic resonant cavity (fig. 2.18c) [164]. This allows the use of a short straight nanowire whilst retaining a high DE (≈30 % [165]). Short SNSPDs have short recovery times (≈500 ps) and may be able to reach count rates of ≈2 GHz [165] and low timing jitters [42]. The high DE is achieved by the placement of the cavity around the nanowire. This cavity can be made by the creation of a Bragg reflector by making local modification to the refractive index of the WG by introducing holes [166] or a grating [167] into the WG. The cavity has to be designed for a specific wavelength which increases the fabrication complexity since low fabrication tolerances are required to obtain the highest DE at the wavelength of photon source. Another possibility is the use of a tapered WG design (fig. 2.18d). This design provides a lot of room for the detector and leaves a lot of freedom to the nanowire design. It allows the integration of complex detector designs as needed for a photon-number resolving detector. The fabrication of a taper can be done straight forward and allows for larger tolerances in the detector placement during fabrication. It enables the integration of detectors with an optimized design to reduce current crowding. Hence, a tapered WG design was chosen, for this thesis that tapers the WG from a width of 600 nm to a width of 10 μm.

2.3.9 Photon-Number Resolution

Photon-number resolution of a detector is the ability to encode the detection of several photons at once in its output signal. This allows the characterization of the emission statistics of a light source by directly measuring their photon-number (PN) statistic (fig. 2.19). This makes photon-number resolving (PNR) detectors useful for a variety of applications in the field of quantum-photonic information technologies and key for LOQC in the KLM-scheme. A SNSPD is generally regarded as a detector that can only count one photon at a time due to its detection mechanism. The growth of the resistive region is facilitated by Joule heating due to the applied bias current. Hence, the pulse amplitude is proportional to the applied bias current and no information can be gained on the energy or number of photons that took part in the detection event. During a detection event, the current is diverted out of the SNSPD, and the nanowire is not biased sufficiently and no further photon can be detected until the SNSPD recovers. PNR resolving can be achieved by the use of multiple SNSPDs [94]. Whilst each detector is still only able to detect one single photon at a time, additional spatially distributed photons can be detected by the other detectors. n-SNSPDs hereby enable n-photon number resolution and, in addition, also spatial resolution. However, for each individual detector a biasing and readout line is required, which significantly increases the complexity of operation and introduces a high thermal load into the cryostat. This can be solved by the use of multiplexing of several detector elements, further regarded as pixels. Time tagged and frequency division multiplexing can be realized using a single RF line. In a time tagged-multiplexing scheme [98] several SNSPDs are connected in series with long delay lines (several hundred micro-meters to milli-meters) between the individual pixels. In dependence on the triggered pixel, the path length the pulses have to travel to be read out differs. The triggered pixel is identified by the timing of the SNSPD pulse relative to either the pulse of a pulsed light source used [98] or in a differential readout by its complementary pulse [99]. When a SNSPD triggers, a positive pulse travels to the side of the detector with a positive voltage potential and a negative pulse travels to the side with a negative voltage potential. When using a differential readout, both sides of the detector are observed and the inter-arrival times of the pulses can be measured. Using the differential readout method in combination with delay lines it was found that the photon-number of the detection event can be extracted out of the pulse shape. However, the interpretation of the pulse shape is difficult and had to be done by hand [99]. In addition, it gets increasingly challenging for large photon numbers. A further drawback is that all pixels are biased in series and one single defective pixel causes the full system to be inoperable. This can be solved by the use of frequency division multiplexing. An embedding of nanowires into resonant circuits (RF-SNSPD), coupled to a transmission line, allows a RF-biasing and readout of multiple pixel using a single RF line [96] [97]. However, RF-SNSPDs require large resonator structures and are sensitive to RF-crosstalk, which prevents the close placement of two RF-SNSPDs next to each other. In addition, this size requirement prevents the placement of several pixel onto a single WG which prevents the use of such type of SNSPD for photon-number resolution on a QPIC.

While photon-number resolution is required for LOQC, spatial resolution is not necessary. Photon-number resolution without spatial resolution can be obtained by the use of amplitude multiplexing. The amplitude multiplexing can be realized by deterministically adding resistors to the nanowire elements (fig. 2.20). Two different concepts can be realized, each of which requires only a single bias and a single readout line for PNR operation and can consequently be used in a standard SNSPD setup without modifications. The first proposed design for amplitude multiplexing using resistors utilizes a parallel design of nanowires and is hence referred to as parallel PNR-SNSPD [100]. A pixel consist of a nanowire section with a series resistors R_S. To obtain PNR, several of these pixels are placed in parallel to each other (fig. 2.20b). The second proposed concept [101] divides a single-nanowire into several sections by a shunting of each section with a parallel resistor R_P. Each pixel consists of a nanowire shunted with a parallel resistor. The pixels are placed in series to each other, hence,

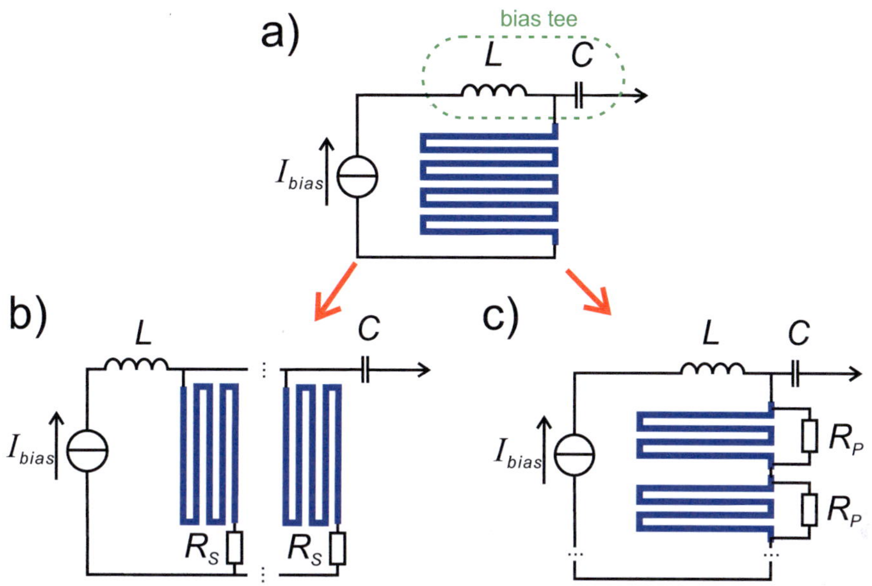

Figure 2.20: a) Electrical representation of a conventional SNSPD with connected bias. The L and C build up the bias tee. The nanowire is colorized blue. b) Electrical representation of a parallel PNR-SNSPD using a series resistor for each detector pixel. c) Electrical representation of a series PNR-SNSPD using parallel resistors for each detector pixel.

this detector type is further regarded as series PNR-SNSPD (fig. 2.20c). In the following section, the operation principle for both types is explained.

For the operation of the parallel PNR-SNSPD, all pixels are biased simultaneously. Ideally the operation of the parallel photon-number-resolving SNSPD is as follows: As the nanowires are superconducting and all R_S have the same value, $N * I_B$ is distributed homogeneously over all pixels and all N-pixels are biased with a bias current I_B. Once a photon triggers a detection event in one pixel, the pixel gets normal conducting and has a resistance in the range of kiloohm and the current gets redistributed. Since the readout is connected through a capacitor acting as a high pass and the L_{kin} of the nanowire and R_S of the remaining pixels act as a low pass, the current gets distributed to the readout and is transformed into a voltage pulse by the impedance of the readout Z (typically $50\,\Omega$): $V_{pulse} \propto I_B \cdot Z$. The triggered pixel is insensitive until it fully recovers. When two photons trigger detection events in two different pixels at the same time, the current out of both pixels gets redistributed and the current distributed to the readout is twice as high. For n-triggered pixels, the pulse voltage gets $V_{pulse} \propto n \cdot I_B \cdot Z$. The height of the output pulse is proportional to the number of absorbed photons. In reality however, a portion of the redistributed current gets distributed to the remaining superconducting pixels. This leakage current $I_{lk}^{(n)}$ increases with the number of triggered pixels n. The bias current in each remaining pixel is increased by $I_{lk}^{(n)}$ which causes an increase of the dark count rate and detection efficiency of the remaining pixels if the SNSPD operates in the probabilistic detection regime. In addition, the maximal number of photons that can be detected is limited by $I_B + I_{lk}^{(n)} \leq I_C$. Hence, $I_{lk}^{(n)}$ significantly influences

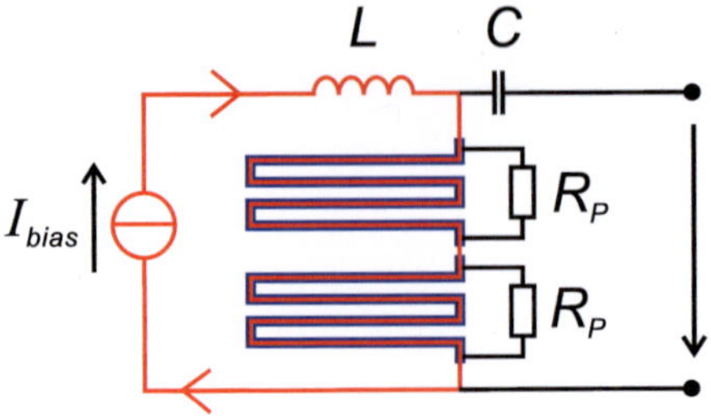

(a) Current flow in idle

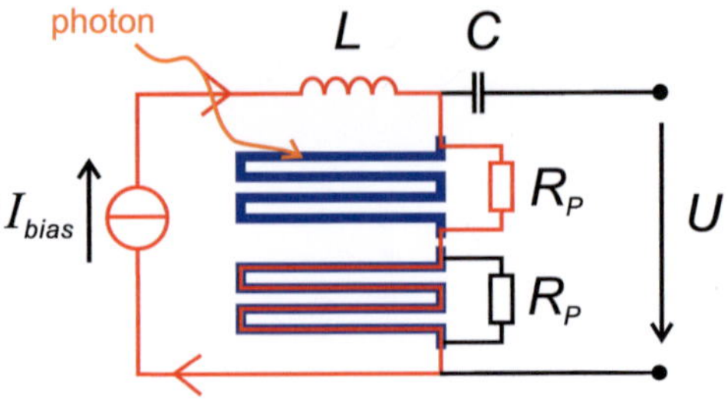

(b) Current flow after 1-photon detection event

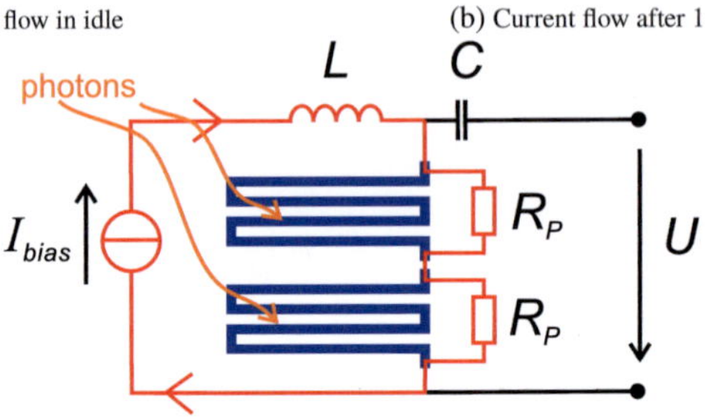

(c) Current flow after 2-photon detection event

Figure 2.21: Illustration of the current flow for the detection of photons in a series PNR-SNSPD.

the performance of the detector. $I_{lk}^{(n)}$ can be reduced by decreasing Z and increasing R_S, L_{kin} and N. A reduction of Z decreases the pulse amplitudes and increases the difficulty to distinguish between amplitude levels. The possible increase of R_S is limited, since it increases the latching probability for the detector [168]. In addition, the nanowires of the individual pixels need to be long enough, to have a sufficient L_{kin} to avoid latching and consequently operate for stable operation. In practices the R_S is limited to $\approx 100\,\Omega$. The problem of the current redistribution limits the PNR capability of the parallel PNR-SNSPD to a resolution of 4 to 6 photons and the efficiency to a few percent, because the bias level has to be chosen appropriately low [100] [40]. The design is robust to a single pixel defect: if the nanowire in one pixel is discontinuous or has a low I_C, the remaining pixels can still be sufficiently biased.

For a series PNR-SNSPD, since all pixels are biased in series, the current out of triggered pixels is not redistributed to the other pixels but over its corresponding shunt resistor instead. The photon detection process for a series PNR-SNSPD is illustrated in fig. 2.21. The nanowire is biased as for a standard SNSPD. Initially the detector is superconducting and the I_B flows through the full nanowire (fig. 2.21a). When a photon triggers a detection event, the resistance of the corresponding pixel is in the range of kΩ. The typical design value of the used shunt resistors is significantly smaller (20 to 100 Ω) and the current gets pushed into R_P (fig. 2.21b). This results in an output voltage pulse: $V_{pulse} \propto I_B \cdot R_P$. The triggered pixel is insensitive until it fully recovers. The redistribution of the bias current into R_P maintains the current flow and all other pixel remain sensitive to incoming photons. When then a second photon triggers a detection event in another pixel (fig. 2.21c), the bias current is diverted into the corresponding R_P and the height of the output pulse is increased by $\propto I_B \cdot R_P$. The output voltage is proportional to the number n of triggered pixels: $V_{pulse} \propto n \cdot I_B \cdot R_P$. In the series

PNR-SNSPD similar to the mechanism in a parallel PNR-SNSPD, a small portion of I_B is distributed to the readout line which can be prevented by the use of a small $R_S < Z$. This redistribution slightly decreases the bias level and, therefore, the DE of remaining pixels. However, this effect is significantly weaker than the current redistribution in the parallel PNR-SNSPD [40]. Due to the small shunt resistor, this detector is self-resetting and intrinsically free from latching. The smaller R_P and the smaller L_{kin}, the faster occurs the redistribution of the current to the shunt resistor and the dead time of the pixel is decreased. However, once the L_{kin} gets to small, the current starts to oscillate between the nanowire and the shunt resistor, which causes oscillating pulses on the readout line. This effect is known as relaxation oscillations. Due to the self resetting, once $I_B > I_C$ relaxation oscillations are observed in the readout line until the nanowire stays normal conducting at an $I_B >> I_C$. Due to this effect, whilst the design is robust if the nanowire in one pixel is discontinuous since the current flow is redistributed through the resistor, once a pixel has a suppressed I_C due to a defect, it causes a number of oscillating pulses and a high amount of noise pulses are observed in the readout. The maximal amount of pixels that can reasonably be implemented using amplitude multiplexing in a series PNR-SNSPD is limited by the number of voltage steps that can be resolved in the readout. This is limited by the dynamic range of the used readout, the noise of the setup and the amplitude jitter of the detector pulse after amplification. So far, PNR-SNSPD with up to 24-pixel have been demonstrated using a series PNR-SNSPD design [169]. By the use of resistors with specific values for each pixel, a spatial resolution can be implemented [106]. However, this significantly limits the amount of pixels that are feasible: for a number of N pixels $\sum_{i=1}^{N} 2^{i-1}$ amplitude levels are needed to encode the spatial information.

For all PNR approaches using SNSPD, the fidelity to record an n-photon event in a single shot measurement is limited by two factors: the probability of photons to hit individual pixels and the detection efficiency of the detector. To record the PN of a multi-photon event, in a single shot measurement with a high fidelity, the number of pixels N needs to be large in comparison to the number of incoming photons n. In addition, the detection efficiency needs to be close to unity, since the probability of detecting a n-photon event is DE^n if all photons hit different pixels. A high accuracy to determine the correct PN of a photon source without a perfect detector can be achieved in a statistical measurement with a sufficient amount of measurements. Taking the intrinsic statistics of the detector into account, the photon statistics of the source can be extracted. The theoretically expected characteristic for a laser photon source, the photon statistics and the detector statistics need to be considered. The lasers have a well-known Poissonian statistics fig. 2.19. The probability of a laser with a mean number of photons per pulse μ to emit n photons is given by:

$$P_\mu(n) = \frac{\mu^n}{n!}\eta^{-\mu}. \tag{2.11}$$

To calculate the probability of a photon triggering a pulse in the detector, several probabilities need to be considered. The probability of correctly detecting n photons out of a pulse that are equally split between N detector elements with an identical detection efficiency η of all elements is [94]:

$$P(n|n) = \frac{N!}{N^n(N-n)!}e^n. \tag{2.12}$$

What is more, a pixel may miss to detect one photon but may detect a second photon or a third photon. By considering the photon correlation in combination with eq. (2.12), the probability $P_\eta^N(n|\mu)$ of detecting n photons from an optical pulse with a Poissonian statistics using a N-element detector is described by [170] [94]:

$$\begin{aligned} P_\eta^N(n|\mu) = \sum_{m=n}^{\infty} \frac{N!}{n!(N-n)!} \frac{(\eta\mu)^m e^{-\eta\mu}}{m!} \times \\ \sum_{j=0}^{n} (-1)^j \frac{n!}{j!(n-j)!} \left[1 - \eta + \frac{(n-j)\eta}{N}\right]^m. \end{aligned} \tag{2.13}$$

For eq. (2.13) to be valid, the detection efficiency of all pixels needs to be identical.

3 AlN Buffer Layer for High-Quality NbN Films on GaAs

To enable efficient SNSPD operation at 4.2 K, the growth of high-quality NbN films on GaAs has been studied. Since the uniform NbN film growth directly on GaAs is challenging [81] [80], an aluminum nitride (AlN) buffer layer is proposed in this chapter as an intermediate layer between GaAs and NbN(fig. 3.1). AlN acts as an efficient diffusion barrier for oxygen ions [171] and potentially also of arsenic, effectively preventing diffusion of substrate components into the superconducting film. Depending on the growth conditions, AlN will grow in a wurzite or a zinc-blende lattice. Grown in a (002) wurzite lattice, the lattice mismatch to NbN is only 8 %. Grown in (100) zinc-blende there is almost no lattice mismatch to NbN [172]. AlN is a high band gap semiconductor (bandgap energy ≈6 eV) and is transparent in the full optical range. Consequently, it can be placed between the SNSPD and a waveguide without blocking photons from reaching the detector. With 380 W/mK [173] AlN has a larger thermal conductivity than GaAs (160 W/mK) [174] at cryogenic temperatures of 4 K. This could be beneficial for superconducting detectors due to a faster energy transfer to the substrate and, consequently, a shorter relaxation time. The improvement of the film quality of NbN and NbTiN films using an AlN buffer layer has already been demonstrated for SiO_2 and soda lime glass substrates [175] but had not yet been demonstrated for GaAs. In addition, to the use as a buffer layer, AlN itself is a very promising candidate for the realization of an integrated photonic chip. It has a significant optical second-order non linearity which makes it suitable as a future material system for photonic integrated circuits with directly integrated single-photon sources [176]. Integrated photonics on AlN have been recently implemented [177] [34].
In this chapter, different parameters for the fabrication of NbN thin films with AlN buffer layers are investigated and evaluated by characterization of electrical and superconducting properties of the thin NbN films. In the first part, the film growth and the reactive sputter process for NbN and AlN is explained along with film characterization and parameter definitions. In the second part, the NbN quality is investigated in dependence of the used NbN and AlN sputter parameters. In the third part, the dielectric properties of sputtered AlN films are

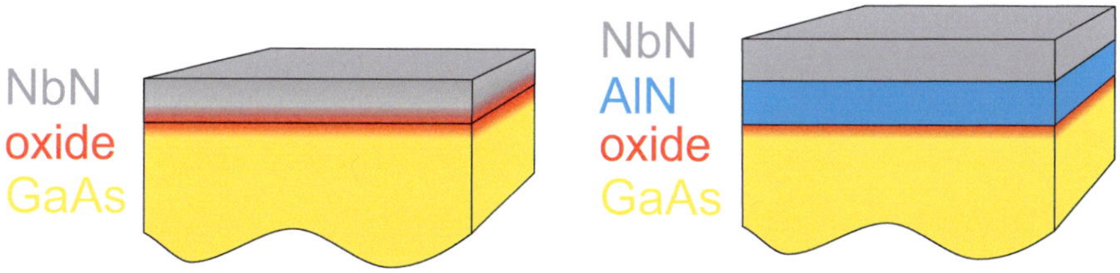

Figure 3.1: Layer structure of NbN on GaAs with and without AlN buffer layer.

investigated using on-chip resonators and the extracted parameters are verified by the on-chip integration of a high-pass filter. Results presented in this chapter are published in part in IEEE Transaction on Applied Superconductivity under the title: "AlN-buffered Superconducting Nanowire Single-Photon Detector on GaAs" [SIS17]. The presented results were achieved with the support of H. Bender [Ben15], S. Barjami [Bar17] and D. Mörixbauer [Mör16].

3.1 DC Reactive Magnetron Sputtering

The NbN as well as the AlN films are deposited using DC reactive magnetron sputtering. Sputtering is a deposition process performed in a vacuum chamber. A low chamber base pressure, typically below 1×10^{-6} mbar, ensures a high purity of the deposited materials. In the chamber, two electrodes are placed in the used system. The cathode is round and has a diameter of approximately 5 cm, the anode forms a ring around the cathode at a distance of a few mm. In a magnetron sputter system a magnet is placed behind the cathode. The north pole of the magnet sits in the middle of the cathode and the south pole forms a ring around the north pole. The field produced by the magnetron increases the ionization rate and allows for a stable plasma in combination with low deposition pressures and voltages. On top of the cathode, a target made out of the to be sputtered material is placed and electrically connected to the cathode. In a DC sputter process allows only conducting targets can be used. A pure Nb target was used during this thesis for sputtering of NbN and a pure Al target was used for sputtering AlN. The substrate is placed opposing the target. The sputtering of the pure target material is performed in a noble gas atmosphere, typically argon, at a pressure in the range of 1×10^{-3} to 1×10^{-2} mbar. When a voltage is applied to the cathode, free electrons are accelerated towards the cathode, ionizing Ar-atoms due to electron-electron collisions. If the ionization rate is higher than the recombination rate of ions, a stable plasma forms and a sufficient amount of ions are available for a stable process. The magnetic field caused by the magnetron catches electrons in a circular path above the target and increases the number of electron-electron collisions, thus increasing the ionization rate. The ionization rate is influenced by the pressure and the applied sputter power. When the Ar-ions hit the target with sufficient energy, material is sputtered off the surface. The target material is scattered inside the chamber and eventually reaches the substrates surface. In order to obtain a high-quality film, the sputtering process must be carefully optimized, since the growth conditions, such as deposition rate, partial and total pressure and deposition temperature, each have a strong influence on the microstructure of the deposited film: on the crystalline orientation, grain sizes, or on internal stress.

3.1.1 NbN Deposition

The I-V characteristic of the Nb sputter process in a pure Ar-atmosphere is depicted in black in fig. 3.2. The Ar pressure was set to $p_{\mathrm{Ar}} = 1.9 \times 10^{-3}$ mbar. The cathode was operated at a constant sputter current. Below 20 mA the plasma is unstable and the discharge voltage

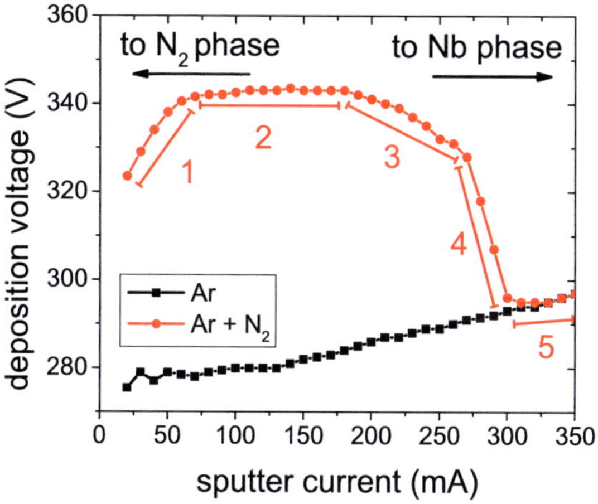

Figure 3.2: Discharge characteristics of the plasma in the NbN deposition system. With a pure Ar atmosphere (black) and in a Ar + N_2 atmosphere (red). (1)-(5) mark different regions in the sputter characteristic.

fluctuates. With an increasing current, the plasma stabilizes and the I-V dependence shows a linear characteristic.

To deposit NbN, N_2 is added to the Ar-atmosphere. The N_2 gets split in the plasma into nitrogen radicals which react with the Nb. The current-voltage characteristic at a total pressure of $p_{\text{total}} = 2.3$ mbar and a partial nitrogen pressure of $p_{N_2} = 4 \times 10^{-4}$ mbar is shown in red in fig. 3.2. The current-voltage dependence can be separated into 5 regions, that are referenced in fig. 3.2 by corresponding digits. For a low sputter current, the resulting deposition rate is generally smaller than the reaction rate of Nb with N. In (1) the full target is covered with NbN and the current-voltage dependence shows a linear characteristic. The voltage is elevated for the NbN compared to the Nb deposition due to a higher secondary electron emission potential [178]. In (2) the deposition rate is outpacing the reaction rate and the characteristic shows a plateau. At (3) the deposition rate is large enough that a significant area on the target is free from NbN and the voltage starts to decrease. At (4) not enough nitrogen is available and almost no NbN is left on the target. The deposition voltage drops until the target is free from NbN and follows the dependence for a pure Ar-atmosphere (5). The stoichiometry of the film can be adjusted by the working point used. A lower sputter current leads to a higher nitrogen content in the film ((1)+(2)). The highest critical temperature (T_C) can generally be found in region (3). Films deposited in the region (4) are more metallic and have a higher Nb content. The largest critical current (I_C) is generally found at slightly higher sputter currents than needed for the highest T_C. Here also the ratio of the measured I_C to the theoretical depairing current I_{dep} is increased [178]. In region (5) the nitrogen content in the chamber is too low and a "dirty" Nb is sputtered, which results in a strong decrease of T_C and I_C. An increase of the nitrogen partial pressure shifts the I-V characteristic towards higher sputter currents and, correspondingly, to larger sputter rates [179].

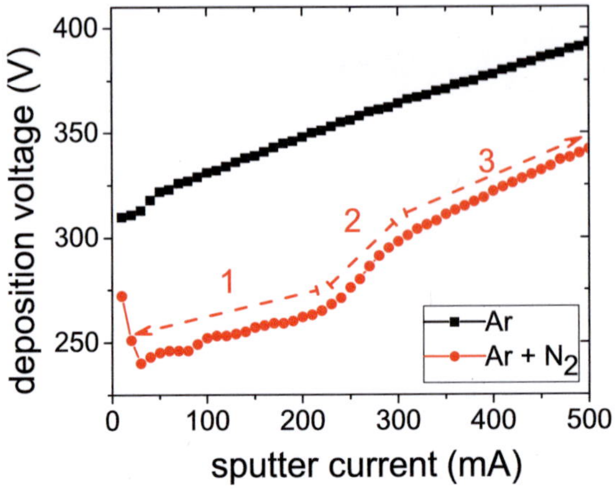

Figure 3.3: Discharge characteristics of the plasma during AlN deposition. With a pure argon atmosphere (black) and a nitrogen and argon atmosphere (red). (1)-(3) mark different regions in the sputter characteristic.

3.1.2 AlN Deposition

The discharge characteristic of the AlN-deposition process used is shown in fig. 3.3. A sputter process in pure Ar-atmosphere (displayed in black) shows a linear characteristic. The curve was recorded at an argon pressure of 2.1×10^{-3} mbar. The cathode was operated with a current control of the sputter current.

To evaluate the discharge characteristic during AlN deposition, N_2 is added at a partial pressure of 0.6×10^{-3} mbar. The I-V characteristic for the reactive process is shown in red in fig. 3.3. In contrast to the NbN deposition as shown in fig. 3.2, the deposition voltage drops with the addition of nitrogen. This can be explained by a larger secondary electron yield of Al with the addition of N_2, which increases the ionization rate [180]. Below 30 mA, the plasma is not stable, so a strong dependence of the deposition voltage on the sputter current can be seen. Three characteristic regions, labeled in fig. 3.3 with digits, can be seen once the current is large enough to allow for a stable plasma. At small sputter currents (1), the deposition rate is smaller than the reaction rate of N_2 with Al, and the full target is covered with AlN. The deposition voltage shows an ohmic rise. When the current is increased, the target starts to get depleted of nitrogen (2) resulting in a stronger rise of the deposition voltage. In the region (3), all nitrogen is used up by the Al and the inclination of the deposition voltage is comparable to the inclination without N_2. Again, the stoichiometry can be adjusted by controlling the sputter current.

3.2 DC Film Characterization and Definitions

The superconducting and normal conducting properties of NbN films were studied using a four-terminal sensing measurement scheme in a liquid helium Dewar at temperatures ranging

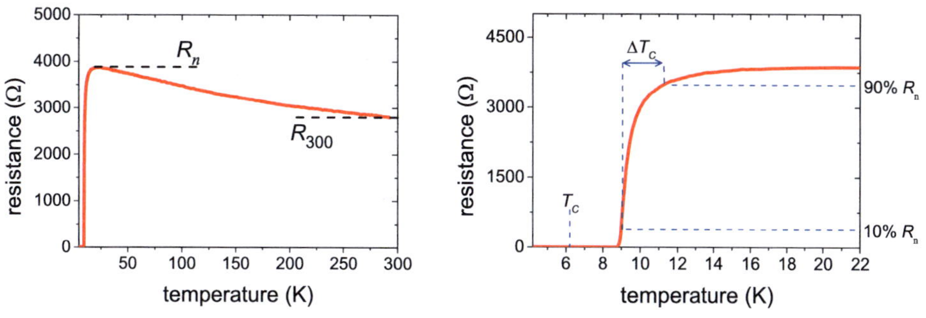

(a) Full dependence

(b) Superconducting transition

Figure 3.4: Temperature dependence of resistance for an 5 nm thick NbN film.

from 4.2 to 300 K. Films were characterized regarding their residual resistivity ρ_n, residual resistance ratio RRR, critical temperature T_C and width of the superconducting transition ΔT_C. The residual resistance and the RRR value can be used to give a qualitative assessment of the defect number in a metallic film. A higher RRR is connected to either a more metallic film or a film with less defects. The T_C gives a qualitative assessment of the superconducting properties of the film, whereas ΔT_C can be used as an indicator for the uniformity of the film. In a non-uniform film, the T_C will vary along the film which results in a large ΔT_C.

The resistance $R(T)$ and, therefore, the resistivity $\rho(T)$ of all studied films increases with decreasing temperature, reaching a maximum value at approximately 20 K (fig. 3.4a). This behavior is typical for thin NbN films [181]. After reaching a maximum, the films start to transition into the superconducting state. In this thesis, the residual resistance R_n is determined as the maximum resistance before its superconducting transition with its corresponding resistivity of ρ_n. The resistance measured at room temperature is further regarded as R_{300} with its corresponding resistivity of ρ_{300}.

Out of a measurement of the resistance at room temperature R_{300} and the normal resistance just before the superconducting transition R_n (see fig. 3.4a). the residual resistance ratio RRR can be calculated as a qualitative factor using an adapted Matthiessen rule:

$$RRR = \frac{R_{300}}{R_n} = \frac{\rho_{\text{phonon}}(300\,\text{K}) + \rho_{\text{electrons}}(300\,\text{K}) + \rho_{\text{defects}} + \rho_{\text{tunneling}}(300\,\text{K})}{\rho_{\text{defects}} + \rho_{\text{tunneling}}(20\,\text{K})}. \tag{3.1}$$

The Matthiessen rule describes the resistivity of a metal as caused by collisions of electrons with phonons, electrons and defects. The resistivity portions, that are caused by collisions with phonons $\rho_{\text{phonon}}(T)$ and electrons $\rho_{\text{electrons}}(T)$ are temperature-dependent and are significantly reduced at lower temperatures [182]. The resistivity at low temperatures is mainly determined by material defects, which is why the RRR value can be used to evaluate the amount of defects in the material. This leads to a positive temperature coefficient for metals. The negative temperature coefficient for thin NbN films cannot be explained by collisions of electrons with

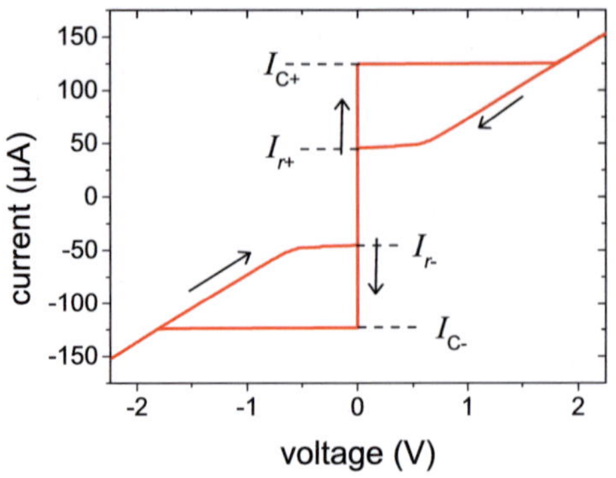

Figure 3.5: Current voltage characteristic of a superconducting film.

phonons, electrons and defects alone. These films have to be viewed not as a smooth film, but as a highly granular film with a large number of grain boundaries, acting as a random array of tunnel junctions with thermally activated tunneling [183]. At larger temperatures, the tunneling probability is higher, because of the higher thermal energy. This can explain a lower resistivity of the film at room temperature. To accommodate for tunneling, $\rho_{\text{tunneling}}(T)$ was added to eq. (3.1). Since the tunnel junctions will act as random array of Josephson junctions [184] in the superconducting state, this effect will not prevent the film from getting superconducting.

T_C was determined (fig. 3.4b) as the temperature at which the resistance drops to 0.1 % of ρ_n during the superconducting transition. The T_C is material-dependent and is strongly influenced by material defects. ΔT_C was determined out of the temperature difference between 10 % and 90 % of ρ_n.

A structuring of bridges out of the NbN film, allows for an accurate measurement of the critical current I_C and the re-trapping current I_r. I_C and I_r can be derived out of a current-voltage characteristic. An example of a current-voltage characteristic for a NbN bridge is shown in fig. 3.5. Starting at a current of 0 μA, the film is superconducting and no voltage drop across the bridge is visible. When the current is increased, no voltage is visible until I_C is reached and the bridge abruptly switches to its normal conducting state. The critical current and corresponding critical current density j_C allow the evaluation of the superconducting properties of the film. If the current is further increased, a resistive dependence is visible. When the current is gradually reduced, the voltage follows a resistive dependence until the bridge switches to its superconducting state. This switchback current is the retrapping current I_r. I_r is defined by the thermal coupling of the superconducting film to the substrate and defines the point where the energy introduced by Joule heating equals the energy reduction by thermal relaxation. The same dependence can be seen for negative currents. The I_r and the corresponding re-trapping current density j_r are connected to the thermal coupling of the film. In combination with ρ_n the thermal coupling B of the film to the substrate can be calculated as [185]:

$$j_r = \sqrt{\left[\frac{B}{4d\rho_n} \cdot (T_C^4 - T_B^4)\right]}, \tag{3.2}$$

where d is the thickness of the film and T_B is the bath temperature, in our case 4.2 K.

3.3 Dependence of NbN Properties on Deposition Conditions

To get a high film quality, different deposition conditions for the NbN sputter process were investigated. The films were sputtered onto a polished GaAs-substrate with and without AlN buffer layer with a (100) crystal orientation. The influences of the substrate temperature during deposition and of the sputter current were investigated. In addition, the influence of the AlN buffer layer was studied for several NbN thicknesses.

3.3.1 Deposition Temperature

For the deposition of NbN thin films on GaAs, three temperature-dependent effects should be taken into account. The first effect is the influence of the temperature on the crystalline growth of thin films. The second effect is connected to the thermal stability of the substrate: elevated temperatures cause the desorption of surface oxides and arsenic from the GaAs-substrate. The third effect is an increased diffusion at higher temperatures, which can lead to an inter-diffusion of tangent layers. For the film growth, higher temperatures result in a higher mobility of the sputtered material on the substrate. This allows the sputtered particles to arrange in a more favorable position on the substrate surface, decreasing the number of lattice defects and increasing the size of individual grains. Depending on temperature and pressure, several zones of film growth are present. In fig. 3.6, several growth zones are shown in dependence of the process pressure and on the normalized substrate temperature. T/T_m here is the substrate temperature during deposition normalized to the melting temperature of the sputtered material. The melting temperature of NbN is 2573 °C.

In zone 1, the mobility of atoms on the substrate surface is low. Atoms stick where they land, resulting in a fine grained porous film structure. In zone 2 surface diffusion occurs with activation energies of 0.1 to 0.3 eV, resulting in a columnar growth of the film. At even higher temperatures, in zone 3, bulk diffusion occurs with activation energies above 0.3 eV resulting in a rough grained equiaxed film structure. Between zone 1 and zone 2 a transition zone, zone T, develops with a smooth surface topology [186] [188]. For the used process pressure at around 2 mtorr, the intermediate zone starts at a substrate temperature of around 70 °C. The film is optimized almost exclusively in zone T.
GaAs has naturally occurring surface oxides with a distinct temperature stability, which causes a significant influence of the deposition temperature on the substrate surface topology [82].

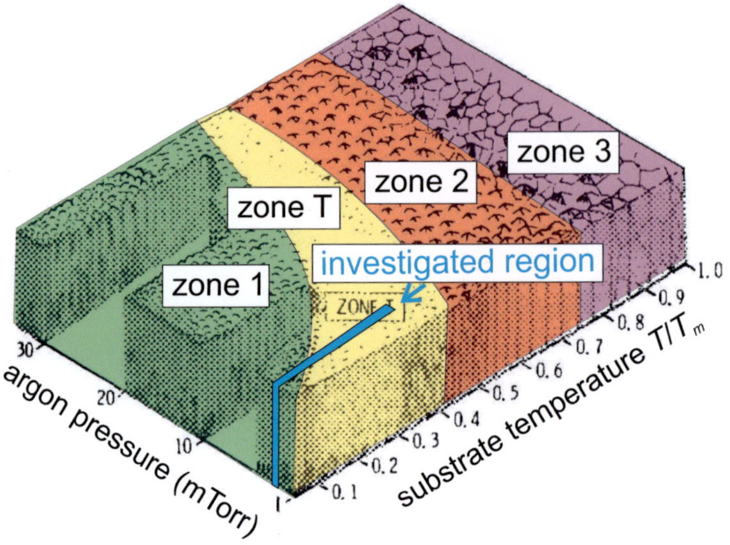

Figure 3.6: Micro-structure zone diagram for sputtered metal films at different temperatures and pressures. T/T_m is the substrate temperature during deposition normalized on the melting temperature of the sputtered material. The investigated temperature region is marked in blue. Reprinted (adapted) from [186] [187] and colorized with the permission from AIP Publishing.

At substrate temperatures of around 150 °C, volatile As-oxides like AsO are desorbed from the surface. With a further increase in temperature, chemical reactions between the GaAs surface and the existing oxides are favored. At approximately 300 °C, the stable As-oxide As_2O_3 reacts with GaAs to the most stable Ga-oxide Ga_2O_3 and arsenic:

$$As_2O_3 + 2\,GaAs \xrightarrow{300\,^\circ C} Ga_2O_3 + 2\,As_2 \quad (\text{or} \quad As_4). \tag{3.3}$$

The arsenic created in this reaction evaporates from the substrate surface. At temperatures around 400 °C, Ga_2O starts to evaporate from the surface. At temperatures of approximately 500 °C, Ga_2O_3 is reduced to the more volatile Ga_2O, which evaporates from the surface:

$$Ga_2O_3 + 4\,Ga \xrightarrow{500\,^\circ C} 3\,Ga_2O. \tag{3.4}$$

The different temperature stability of the present oxides leads to the growth of surface pits on GaAs. Stable oxides mask the surface, whilst volatile oxides and arsenic are desorbed [82]. In addition, diffusion processes are accelerated at high temperatures which decreases the film's quality by inter-diffusion of O and As atoms into the NbN and of Nb into the GaAs [80].
In summary, an elevated temperature is preferred for the crystalline film growth, since a higher surface mobility fosters epitaxial growth and reduces the number of surface defects. On the other hand, the surface of the substrate roughens up and a higher temperature leads to an

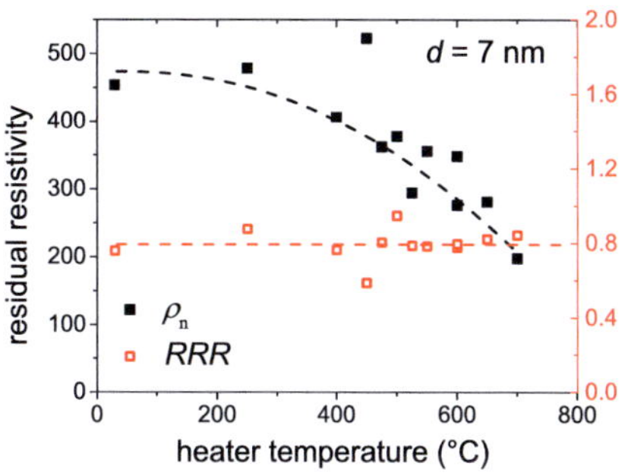

(a) Residual resistivity and RRR

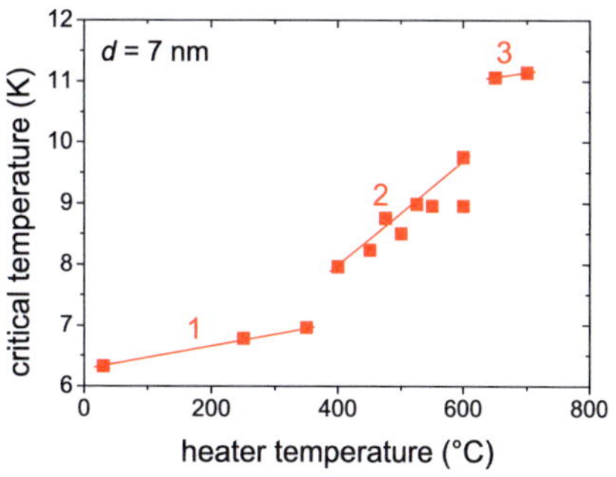

(b) Critical temperature

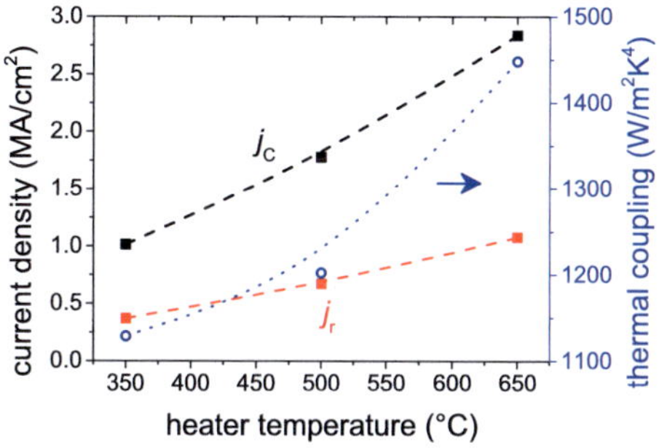

(c) Critical current density j_C and re-trapping current j_r.

Figure 3.7: Superconducting properties of NbN films in dependence of the deposition temperature.

increased diffusion of oxygen and arsenic into the NbN film decreasing the quality of the sputtered film.
To find the optimal deposition temperature of NbN films, the substrate was placed on a heater. Films were deposited on pure GaAs (100) substrates at heater temperatures of 30 to 700 °C. In addition, a sapphire substrate was placed along the GaAs substrate, which was later used to control the thickness of the grown film. The substrates were positioned on top of a copper substrate holder, which was placed on top of the heater without the use of thermal glue. The presence of two thermal interfaces with a not optimal thermal coupling causes the substrate temperature to be significantly lower than the heater temperature. The deposition temperature was reached by increasing the heater temperature from room temperature to the deposition temperature at a rate of 20 °C/min. The substrates were kept approximately 30 min at the chosen deposition temperature prior to the film growth in order to ensure the substrate to reach the state of thermal equilibrium. 7 nm of NbN was sputtered at a current of 245 mA at partial pressures of $p_{\mathrm{Ar}} = 1.9 \times 10^{-3}$ mbar and $p_{\mathrm{N_2}} = 0.4 \times 10^{-3}$ mbar. The resulting deposition rate is 0.11 nm/s. After growth, the film thickness was controlled on the sapphire substrate after patterning stripes using photo-lithography and reactive ion etching (RIE). Profilometry was used to measure the thickness.

The samples were characterized regarding their resistive and superconducting properties. The residual resistivity decreases with increasing heater temperature t_{heater} while the residual resistance ratio RRR shows no dependence on the deposition temperature (fig. 3.7a). By increasing t_{heater}, T_{C} increases by a total of 5 K to a T_{C} of 11.2 K (fig. 3.7b). The dependence of the T_{C} on the t_{heater} can be categorized into three temperature regions, with jumps between the individual regions. In region (1) where the temperature rises from 30 to 350 °C, the T_{C} shows a moderate growth on temperature. In region (2) in the range from 400 to 600 °C, this growth is enhanced and in region (3) the growth is reduced. The increase of T_{C} and decrease of ρ_n at higher heater temperatures can be explained by a more ordered crystal growth of the film due to the higher mobility of atoms on the substrate surface, reducing the amount of defects in the film. Since RRR is expected to increase in case of a lower amount of defects, the observed temperature-independence of RRR is unexpected. The temperature independence of the RRR can be explained by a shunting effect of the GaAs substrate at high temperatures which dominates the resistance-temperature dependence. At room temperature, the resistivity of GaAs is similar to the resistivity of the normal conducting NbN film. While cooling down, the conductivity of GaAs decreases exponentially with decreasing temperature and is significantly lower than the conductivity of the NbN film at 20 K. Hence, the RRR is not suitable for an evaluation of the film properties for NbN films directly on GaAs. The three temperature regions of T_{C} found can be explained by changes in the substrate topology due to substrate heating. In region (1), most of the native oxides on the GaAs surface are still present. The better crystal growth is counteracted by the interdiffusion of arsenic and oxygen into the NbN film. With increasing temperature, As-oxides start to evaporate from the surface. The As desorption is enhanced by the (eq. (3.3)) decomposition of As_2O_3 taking place at substrate temperatures of approximately 300 °C, which fits very well with the jump from region (1) to region (2) at t_{heater} = 400 °C. The stronger T_{C} growth in region (2) can be explained by a lower amount of remaining surface oxides due to the 30 min annealing. The transition from region (2) to region (3) at t_{heater} = 650 °C fits well to the decomposition of Ga_2O_3 (eq. (3.4)) taking place at 500 °C, which leads to the desorption of the stable Ga-oxides, further decreasing the amount of oxygen on the surface. However, the roughness of the surface grows and the T_{C} growth in region (3) is reduced. Out of the assignment of the data to different surface desorption processes, the substrate temperature can be estimated to ≈500 °C at a heater temperature of 650 °C.

The critical current density j_{C} and the re-trapping current density j_r of the NbN films was investigated using micrometer-sized bridges (fig. 3.7c). With increasing t_{heater} from 350 to 650 °C the j_{C} and j_r almost tripled to a j_{C} of 2.8 MA/cm^2 and a j_r of 1.1 MA/cm^2, respectively. A heating of the substrate during the deposition process significantly improves the thermal coupling B by 25 % (fig. 3.7c).

In conclusion, it can be observed that a heating of the substrate during deposition significantly increases T_{C} and the thermal coupling of the NbN film on GaAs. The benefits of an improved crystalline growth outweighs the effects of the temperature stability of the substrate in the investigated range with heater temperatures ranging from room temperature to 700 °C and a further increase in T_{C} and j_{C} is expected for even higher temperatures. However, for the growth of films with self assembled QDs, the heater temperature needs to be significantly lower than the annealing temperature during QD-fabrication (610 °C) in order to prevent a degradation of QDs.

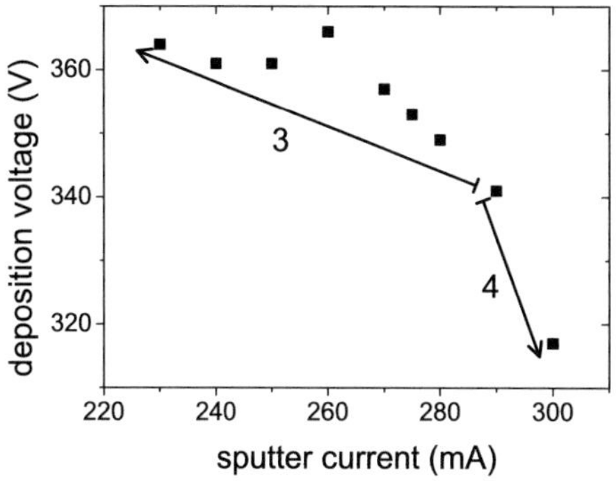

(a) Deposition voltage

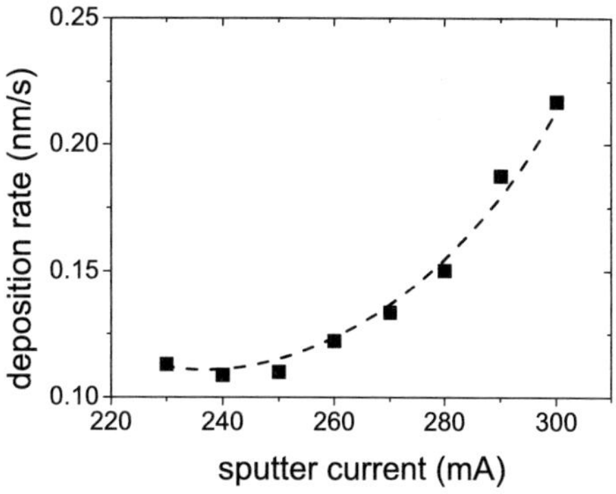

(b) Deposition rate

Figure 3.8: Deposition voltage and deposition rate in dependence on the sputter current in the NbN deposition system.

3.3.2 Sputter Current

A variation of the sputter current for the NbN deposition allows for an optimization of superconducting properties by changing the film stoichiometry and the crystal growth. The film growth is investigated on GaAs (100) substrate with and without AlN buffer layer to find the deposition condition for which the highest T_C and smallest ΔT_C can be reached. The influence of the AlN buffer layer is evaluated by directly comparing films with and without a buffer layer. The AlN buffer layer was sputtered using DC reactive magnetron sputtering out of an Al-target in a nitrogen and argon atmosphere at a heater temperature of 600 °C. The substrates were placed onto the heater without the use of thermal glue and were kept at 600 °C approximately 30 min prior to the AlN deposition. 12 nm of AlN was grown at a sputter current of 160 mA, a pressure of 2.7 mbar and a p_{Ar}:p_{N_2} ratio of 3.6:1 resulting in a deposition rate of 0.08 nm/s. All AlN films were deposited ex situ to the NbN film in a single sputter process for all investigated substrates to yield the best possible comparability.

NbN films with d_{NbN} = 5.5 nm were deposited onto buffered and raw GaAs substrates simultaneously. The films were deposited at a heater temperature t_{heater} of 600 °C. Prior to the NbN deposition, the substrates were kept at the deposition temperature for approximately 30 min to allow for thermalization. The NbN was sputtered at a pressure of p_{total} = 3.3 mbar at a p_{Ar}:p_{N_2} ratio of 2:1. The sputter current was optimized in the transition between regions (3) and (4) of the current-voltage characteristic of the NbN sputter system (see section 3.1) from 230 to 300 mA. The corresponding deposition voltages are shown in fig. 3.8a. As expected, the deposition rate raises with an increasing sputter current (fig. 3.8b).

T_C (fig. 3.9a) increases with increasing sputter current until it reaches a maximum of 12 K at 275 mA and then decreases strongly. The T_C is increased by the use of an AlN buffer layer by ≈2 K over the full investigated range of sputter currents. A T_C of 12 K is a significant improvement over previously reported values for ≈5 nm thick NbN films (8.2 K [81]; 8.7 K [80]; 9.4 K [189]. All these values were adapted to the 0.1 % ρ_n criterion used for the determination of T_C during this thesis). The ΔT_C decreases with an increasing sputter current

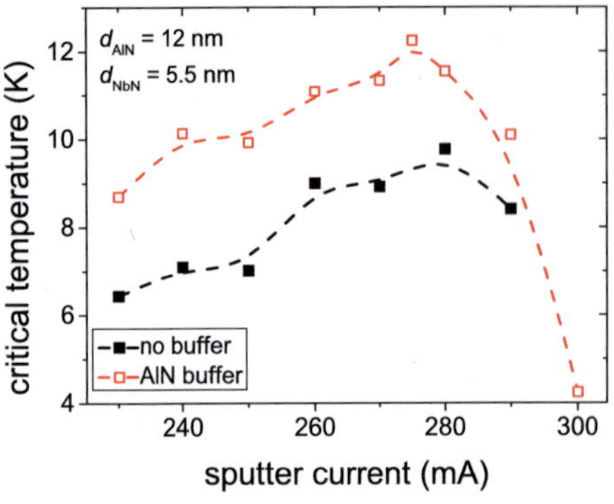

(a) Critical temperature

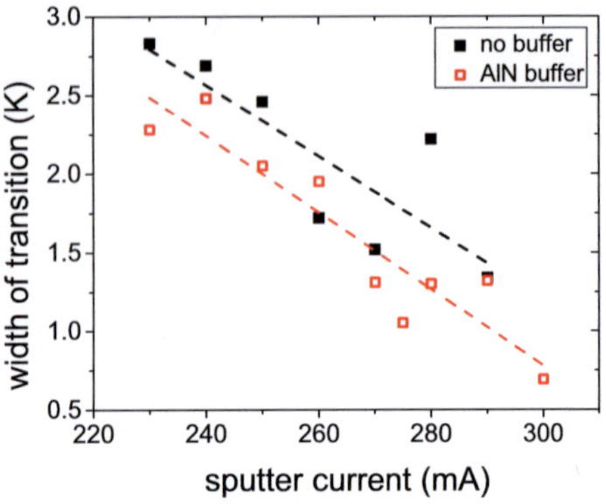

(b) Width of the superconducting transition

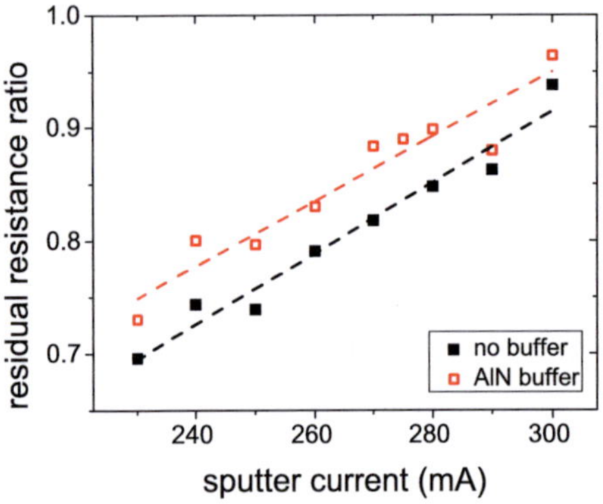

(c) Residual resistivity

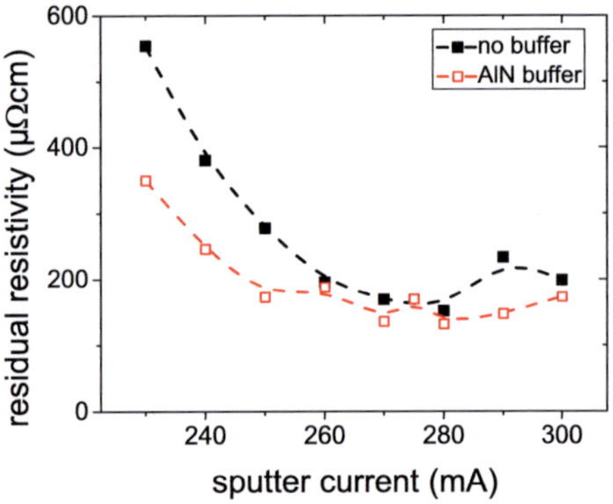

(d) Residual resistance

Figure 3.9: NbN film properties in dependence of the sputter current and influence of a AlN buffer layer. Films without buffer layer are displayed in black, films with a 12 nm AlN buffer layer in red. The dashed lines are to guide the eye.

(fig. 3.9b) and is decreased by ≈0.4 K due to the buffer layer. *RRR* increases with an increase of the sputter current. An by approximately 0.06 larger *RRR* value is measured for NbN films on the buffered substrate. The residual resistivity is lower for NbN on a AlN buffer layer. The residual resitivity decreases for an increasing sputter current until 280 mA and then slightly increases for NbN films on both substrates. Overall, the AlN buffer layer improves the superconducting properties of the NbN film. The largest T_C and the lowest residual resistance is found to be at the end of sputter region (3) just before the transition to region (4). Since this "optimal" point is found for the same current independent of the substrate, this indicates an optimal stoichiometry of the film. The decrease of ΔT indicates a better homogeneity of the film at higher sputter currents. The rise of the *RRR* can be attributed to a more metallic film due to a shift of the stoichiometry towards a higher Nb content and lower number of defects at higher sputter currents.

3.3.3 NbN with Different Thickness

The superconducting properties of the NbN film get suppressed on its surface because of oxidation and on the substrate-NbN interface due to interdiffusion. This normal conducting layer in close proximity to the superconducting layer reduces the effective thickness of the superconducting film and further suppresses the critical temperature. This suppression of superconductivity is known as the superconducting proximity effect [83] [84] [85]. The influence of the superconducting proximity effect is evaluated by changing the thickness of the NbN film since the thickness of the film with suppressed superconductivity stays constant, whilst the thickness of the superconducting core increases with a larger film thickness. Films grown for this section were grown at a heater temperature of 500 °C. The main part of this section is published in [SIS17]. Prior to deposition, the substrate was precleaned in a low power (20 W) RF sputtering process in Ar-atmosphere at a pressure of 1×10^{-2} mbar. The AlN deposition was performed in-situ with the precleaning process at p_{total} of 2.4×10^{-3} mbar at a ratio of $p_{\mathrm{Ar}} : p_{\mathrm{N_2}}$ of 3 : 1. 10 nm of AlN were deposited at a rate of 0.11 nm/s. The NbN films were sputtered simultaneously on a substrate with and without buffer layer, ex situ to the AlN film without additional precleaning. The deposition was performed at p_{total} of 4×10^{-3} mbar at a ratio of $p_{\mathrm{Ar}} : p_{\mathrm{N_2}}$ of 3 : 1, at a deposition rate of 11 nm/s. NbN films were deposited at thicknesses from 3.3 to 20 nm. The films were patterned into single-bridge structures of widths from 1 to 10 µm and a length of 20 µm each.

The presented results are averaged over 10 samples for each individual thickness (fig. 3.10a). The residual resistivity increases with decreasing thickness. For buffered films, ρ_n is 15 % smaller than for films on bare GaAs. *RRR* decreases with decreasing d. For the NbN film with a buffer layer, the *RRR* value is increased by 20 % (fig. 3.10a). T_{C} decreases with decreasing thickness for both substrates and is larger for buffered films in the full range (fig. 3.10b). In addition, to a larger T_{C}, the variation of T_{C} for the measured bridges is two times smaller in the buffered case (0.1 K). This indicates a better uniformity of the NbN film in the buffered case over the full area of the chip (100 mm^2). With increasing thickness ΔT_{C} decreases, but no influence of the substrate can be seen (fig. 3.10b). j_{C} and j_r decrease with decreasing thickness (fig. 3.10c). The re-trapping current is similar for both substrates, nevertheless, the critical current is increased by 73 % by the buffer layer. The calculated thermal coupling coefficient is almost twice as high for thick NbN films without a buffer layer. The coupling decreases for $d_{\mathrm{NbN}} < 10$ nm (fig. 3.10d) steeper without a buffer layer and is only 14 % smaller for the buffered sample for $d_{\mathrm{NbN}} < 5$ nm. The thinnest film of 3.3 nm shows a larger $T_{\mathrm{C}}, j_{\mathrm{C}}, B$ and a smaller resistivity than expected in comparison to thicker films on both substrates. The $T_{\mathrm{C}}(d)$ dependence (fig. 3.10b, solid line) was fitted with $T_{\mathrm{C}}(d)$ as predicted by Cooper [83] [190]:

$$T_{\mathrm{C}}(d) = T_{\mathrm{C}}(\infty) \cdot \exp \frac{\ell_S - 2\Delta d}{NV(d - 2\Delta d)}, \tag{3.5}$$

where NV is the bulk interaction potential and ℓ_S the electron mean-free path in the superconductor. The value of ℓ_S was chosen according to [131] to 0.8 nm for the fit. The critical temperature for the bulk film $T_{\mathrm{C}}(\infty)$ and the thickness of a layer with suppressed supercon-

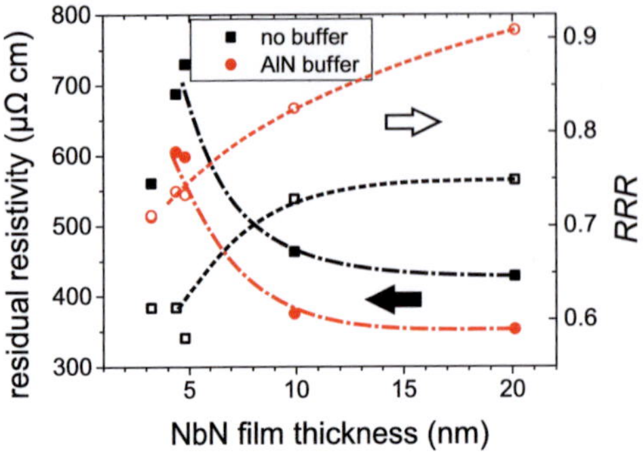

(a) residual resistivity and *RRR* ©2017 IEEE [SIS17]

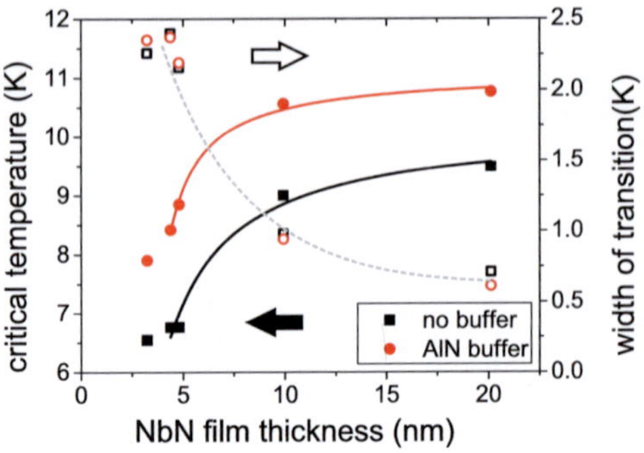

(b) Critical temperature and transition width, T_C was fitted by eq. (3.5) ©2017 IEEE [SIS17]

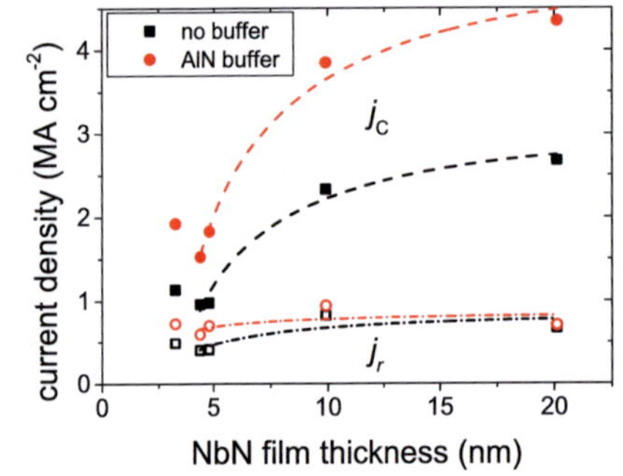

(c) Critical and retrapping current density ©2017 IEEE [SIS17]

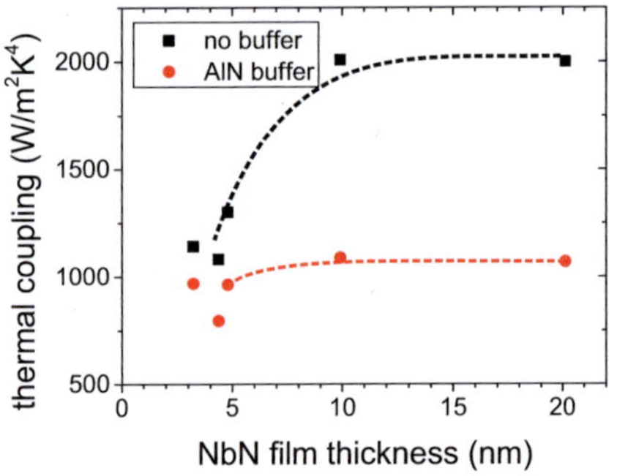

(d) Thermal coupling coefficient

Figure 3.10: Dependence of superconducting and electrical properties on thickness of the NbN film. Films without buffer layer are displayed in black, with an AlN buffer layer in red. The dashed lines are to guide the eye, solid lines are fits to the data with eq. (3.5). The closed symbols in a) and b) correspond to the left Y-axis and the open symbols to the right Y-axis of the figure

ductivity Δd can be extracted out of fit. The fit reveals a $T_C(\infty)$ of 10.2 K without and 11.5 K with AlN buffer. The *NV* was determined to 1.5 out of the fit. The extracted Δd corresponds to 0.86 nm for the unbuffered and 0.68 nm for the buffered film. Hence, the thickness of the film with suppressed superconductivity is reduced by the buffer layer. AlN-buffered NbN films have a larger critical temperature, lower resistivity and a larger *RRR*. Correspondingly the j_C at a temperature of 4.2 K is increased. However, no influence of the buffer layer on ΔT_C can be seen. These results can be explained by the AlN buffer layer having a polycrystalline structure or rather amorphous structure in contrast to an ideal mono-crystalline structure with well-defined orientation. This AlN crystal structure may cause the NbN growth to also favor a granular polycrystalline to amorphous structure in contrast to a more epitaxial growth of NbN films onto heated sapphire substrates [131]. The reduced thermal coupling coefficient of the buffered substrate also supports a polycrystalline or amorphous nature of the AlN film, since such a structure has a weakened thermal conductivity in comparison to an epitaxial film. This indicates that the improvement of the NbN film quality is connected to the AlN buffer layer acting as a diffusion stopper. It prevents oxygen and arsenic diffusion into the NbN and

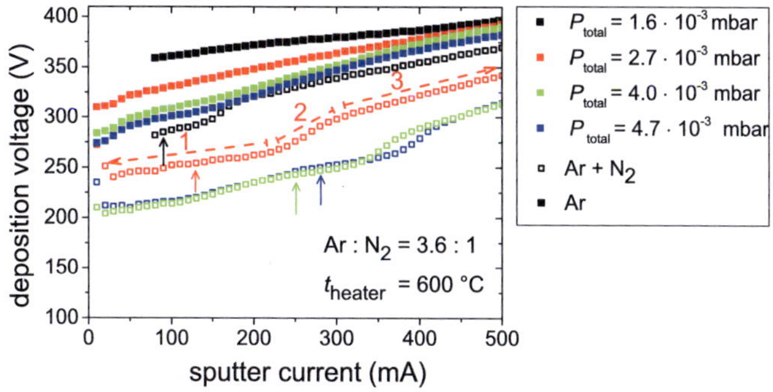

Figure 3.11: Discharge characteristics of the AlN deposition system in dependence on the process pressure. The characteristic of the AlN deposition consist of a deposition regime with excess nitrogen (1), a transition region (2), and a region of excess aluminum (3). Arrows mark the working points with identical relative nitrogen consumption that were used for evaluation.

the formation of non-superconducting Nb – As compounds and oxidation at the NbN substrate interface. The NbN film can be considered as a superconducting body that is sandwiched between normal conducting layers with high resistivity which is caused by surface oxidation. The increase of T_C and consequently j_C of the film can be explained in the context of the superconducting proximity effect, since the thickness of the superconducting layer is reduced by a smaller Δd for the buffered NbN film. This further explains the lower resistivity of the buffered film as the lower resistive core remains thicker. Furthermore, the lower defect density in the thicker core in comparison to the edges explains the increase of the measured RRR.

3.4 Dependence of NbN Properties on AlN Deposition Conditions

As seen in the previous section, the AlN buffer significantly improves the superconducting properties of the NbN film. Furthermore, it was concluded that the crystalline structure of the AlN buffer layer has a significant influence on the growth and, therefore, the quality of the NbN film. For a further improvement of growth conditions, an investigation of the influence of AlN growth conditions on the quality of the NbN films was performed. In this section, all NbN films were deposited at a p_{total} of 2.4×10^{-3} mbar at a ratio of $p_{\text{Ar}} : p_{\text{N}_2}$ of 3 : 1 at a sputter current of 260 mA at a deposition rate of 0.13 nm/s.

3.4.1 Working Pressure

The working pressure has a considerable influence on the growth of the film during sputtering. It influences the free path length of the ions and the free path length of the sputtered material during the process. An adjustment of the working pressure allows an adjustment of the deposition rate while maintaining the stoichiometry of the film.

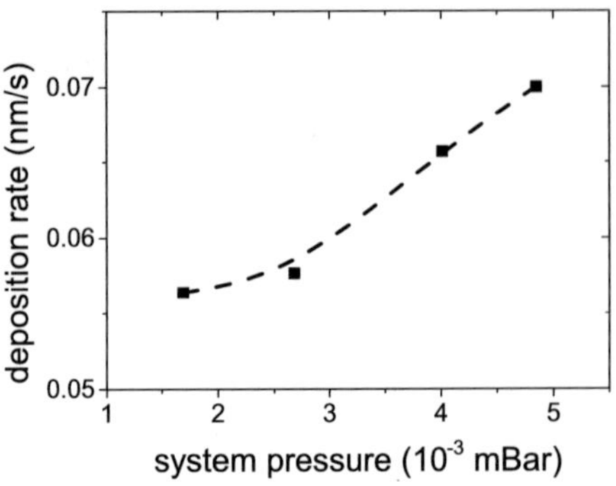

Figure 3.12: Deposition rate in dependence on the pressure at the used working points. The dashed line is to guide the eye.

The NbN film quality was investigated for four different AlN deposition pressures from $p_{\text{total}} = 1.6 \times 10^{-3}$ mbar to 4.7×10^{-3} mbar. The partial pressures were adjusted to a ratio p_{Ar}:p_{N_2} of 3.6:1. The corresponding voltage-current dependencies are shown in fig. 3.11. For the lowest pressure, the dependence starts at 70 mA since the plasma was not stable for lower currents. In region (1) the resulting films are transparent and start to get a metallic look in region (2) indicating an excess of aluminum in the film until it gets fully metallic in region (3). Since for photonic integrated circuits a transparent buffer layer is a necessity, the investigation of AlN films was focused on region (1). For the different working pressures, the sputter current was adjusted to the same relative nitrogen consumption ΔN_2 of 69 % to maintain the stoichiometry for all films. The nitrogen consumption was defined as the amount of nitrogen that is consumed during deposition and was calculated as:

$$\Delta N_2 = 1 - \frac{p_{\text{total}}(I_{sp}) - p_{\text{Ar}}(0)}{p_{\text{N}_2}(0)}. \tag{3.6}$$

$p_{\text{total}}(I_{sp})$ is the total pressure at an applied sputter current (I_{sp}), $p_{\text{Ar}}(0)$ and $p_{\text{N}_2}(0)$ are the partial pressures with no applied sputter current. The corresponding working points are marked by arrows in fig. 3.11. 5 nm of AlN was deposited at each working pressure. On all AlN samples, 5 nm of NbN was deposited simultaneously ex situ to the AlN deposition. Both depositions were performed at a heater temperature of 600 °C.

The deposition rates of AlN increase with an increasing working pressure (fig. 3.12). The T_C and RRR rise for a lower pressure until they drop at the lowest pressure of 1.6 mbar (fig. 3.13). The ΔT_C decreases towards a lower pressure until it rises at the lowest pressure of 1.6 mbar. This can be explained by the lower deposition rate and the energy of particles: a lower deposition rate gives arriving particles more time to arrange in favorable positions in the film before they are buried by new incoming particles. For a lower process pressure, less energy is lost in collisions with other particles. A higher energy on the substrate results in a larger surface mobility of the sputtered material and a more crystalline film growth is

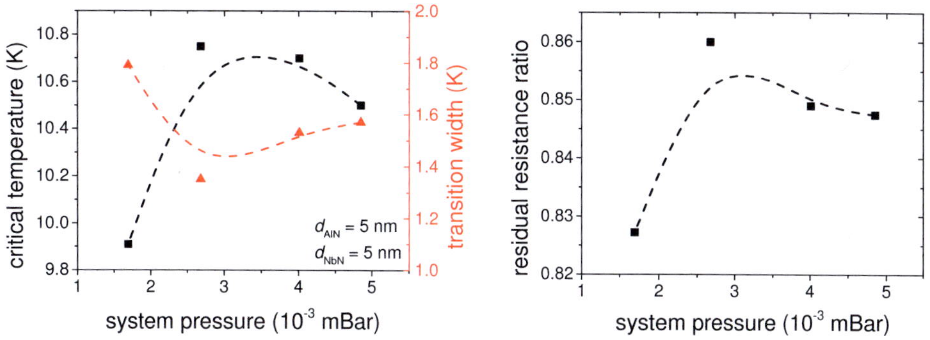

(a) T_C (black) and ΔT_C (red)

(b) RRR

Figure 3.13: Influence of the AlN deposition pressure on the superconducting properties of the NbN layer. To keep the (AlN) stoichiometry constant the nitrogen consumption was kept constant at 69 %. The dashed lines are a b-spline interpolation of the data to guide the eye.

favored. The lower T_C, RRR and higher ΔT_C at the lowest pressure setting can be explained by a inhomogenous film growth of the AlN film due to the plasma not being perfectly stable at the used pressure with the used sputter current. Essentially, the pressure needs to be as low as possible while maintaining a stable plasma during deposition. The optimal deposition pressure for the AlN deposition to reach a high T_C, a low T_C and a high RRR of NbN films is 2.7 mbar in the available sputter system and was used for further AlN optimization.

3.4.2 Sputter Current

Since the sputter current has a significant influence on the growth conditions and stoichiometry of the film, the influence of the AlN sputter current on the electric and superconducting properties of the NbN film was investigated. 5 nm of AlN were deposited for a sputter current from 100 to 300 mA. The corresponding AlN deposition rates are shown in fig. 3.14. The deposition rate increases for an increasing sputter current due to the increased sputter power. A 4.5 nm thick NbN film was deposited onto the AlN, to evaluate the benefit for NbN films.

For an increasing I_{sp} from 160 to 180 mA, the T_C increases then starts to decrease until it levels out at approximately 250 mA, whilst ΔT_C stays constant (fig. 3.15a). RRR slightly increases, however, the values for RRR have a large variation for the investigated samples, which prevents the extraction of a clear dependence (fig. 3.15b). The increase of the RRR towards larger sputter currents is not necessarily connected to lower defects in the NbN film but may also be caused by the change of the AlN stoichiometry towards a higher Al-content, which increases the conductivity of AlN and may lead to a resistive shunting of the NbN film.

To see if the change in sputter rate and stoichiometry influences the AlN surface structure, the roughness of the AlN film was investigated by scanning an area of 5 µm × 5 µm of the samples surface with an atomic force microscope (AFM) and calculating the root mean square deviation of the measured height for all data points. The measurement was performed for

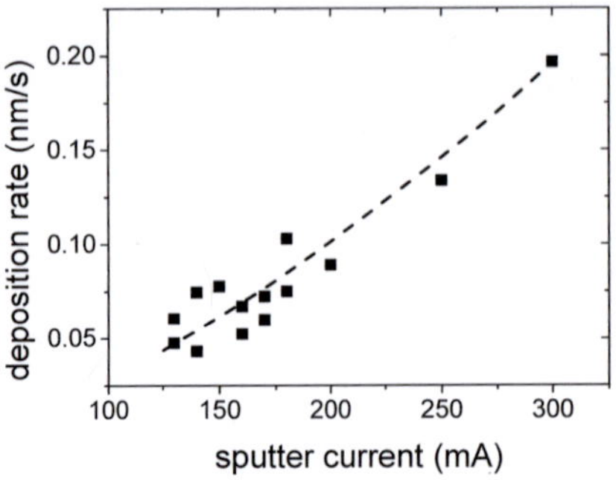

Figure 3.14: Dependence of the AlN deposition rate on the sputter current. The dashed line is to guide the eye

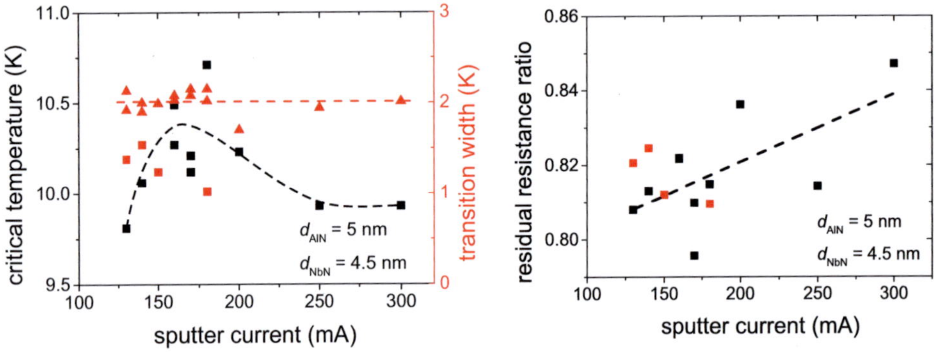

(a) Critical temperature

(b) *RRR*

Figure 3.15: Influence of the AlN sputter current on superconducting and electrical properties of the NbN layer. Dashed lines are to guide the eye.

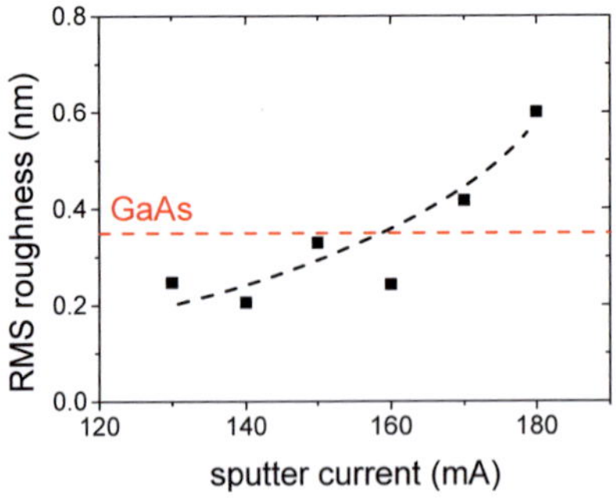

Figure 3.16: Dependence of the AlN film roughness at different sputter current. The measured roughness of a plain GaAs(100) is drawn as a dashed red line. The black dashed line is a guide to the eye for the AlN roughness dependence.

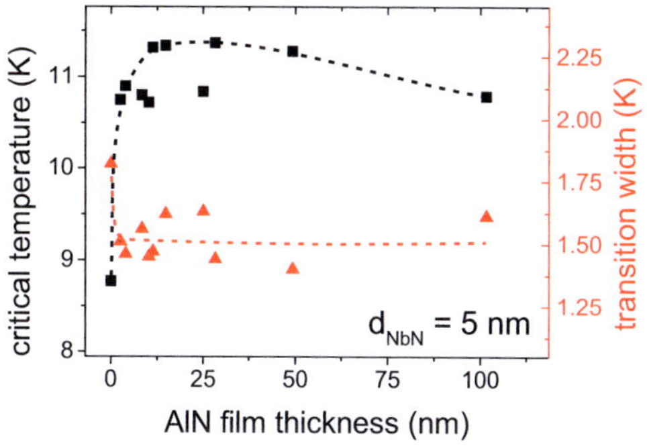

(a) Critical temperature

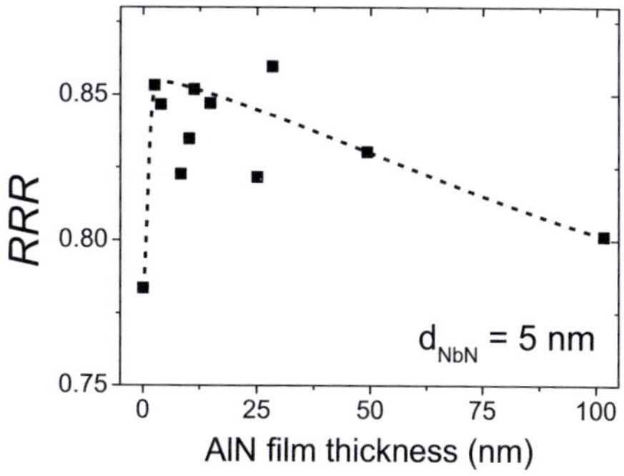

(b) *RRR*

Figure 3.17: Dependence of NbN film properties on the thickness of AlN. The dashed lines are a guide to the eye.

samples sputtered at currents from 130 to 180 mA and compared to the roughness of the GaAs(100) substrate prior to AlN deposition (fig. 3.16). For small currents, the roughness of the AlN films is lower than the roughness of the plain substrate but increases with increasing sputter current and is larger than the GaAs surface roughness for currents larger than ≈165 mA. In total the highest T_C for NbN, and hence the optimal sputter current for AlN, is found at 160 mA.

3.4.3 AlN Thickness

The quality of the NbN was investigated for a variation of the AlN thickness (d_{AlN}). The aim of this variation in thickness was to find the thinnest possible buffer layer for a high-quality NbN film. The thickness should be as small as possible to enable the best coupling of the evanescent WG field to the detector in case of a photonic circuit. However, if the thickness is too thin, the AlN will not act as an efficient diffusion barrier so an optimal thickness has to be found.
AlN buffer layers with thicknesses of 2.5 to 100 nm were deposited at the previously optimized pressure and sputter current. A 5 nm thick NbN film was used for evaluation.

The resulting T_C and ΔT_C are shown in fig. 3.17a. The T_C increase by the buffer layer is approximately 2 K already for the thinnest investigated buffer layer with a thickness of 2.5 nm. The T_C increases until a thickness of 12 nm is reached, stays almost constant until d_{AlN} = 50 nm and then slightly decreases towards 100 nm. ΔT_C is decreased by the presence of the AlN film and does not show a clear dependence on thickness. The *RRR* (fig. 3.17b) is increased by the AlN buffer layer and decreases with increasing AlN film thickness.

The surface roughness of the AlN film (fig. 3.18) is lower than the roughness of the GaAs substrate for AlN film thicknesses below 6 nm. For larger thicknesses, the roughness increases beyond the roughness of the GaAs surface and further increases with increasing thickness. The strong increase in T_C and *RRR* and the decrease in ΔT_C already for very thin AlN films can be explained by the buffer layer acting as a diffusion stopper layer and a change in the

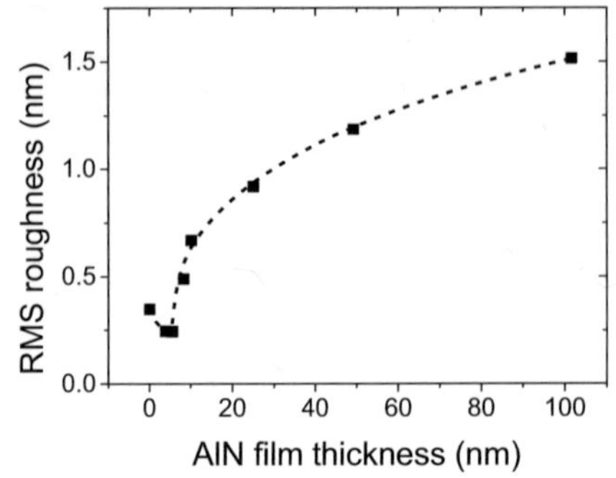

Figure 3.18: Surface roughness of the AlN film at different thicknesses. The dashed line is a guide to the eye.

surface roughness. Even the thinnest tested buffer layer of 2.5 nm thickness consisting of approximately 5 mono layers of AlN may efficiently reduce diffusion while also reducing surface roughness, leading to a significant increase in T_C and RRR. The T_C is further increased until $d_{AlN} \approx 12\,\mathrm{nm}$ since the buffer layer gets more impermeable until it stops all diffusion. The decrease of T_C and RRR towards lager thicknesses can be explained by the increasing roughness for larger AlN film thicknesses. In contrast, while the ΔT_C is reduced by the presence of the buffer layer, no dependence on the AlN thickness is visible. The best conditions of NbN film growth is found for a AlN thickness of 12 nm. The T_C of 10.9 K of the NbN film at a AlN thickness of 100 nm is large enough to indicate that detector operation at 4.2 K is feasible even for thick AlN films. This provides the opportunity of the fabricating AlN-based photonic integrated circuits on sputtered AlN films.

3.5 Dielectric Properties of the AlN Film

Whilst the quality of the NbN film can be assessed directly by the measurement of its electrical properties, this is not possible for an isolating film. To directly investigate the quality of the fabricated AlN films, their dielectric properties were evaluated. A high-quality AlN film is not only interesting as a buffer layer but also as a functional layer for integrated photonic circuits. It could further be used as a dielectric for capacitors integrated with SNSPDs to build up an integrated bias tee. Integrating a bias tee onto the same chip as the detector would reduce routing complexity and allow the reduction of thermal noise since it is cooled down to the same temperature as the SNSPD. An integrated bias tee would allow the individual readout for a larger number of detectors in a cryostat with limited available space. As demonstrated by Hofherr [191], a large enough on-chip capacity cannot be reached by means of a planar inter-digital capacitor whilst adhering to the space requirements. A solution to this is the use of a thin film plate capacitor with AlN as a dielectric. To evaluate the quality of the NbN film, a resonator design was developed that allows the characterization of the film's dielectric constant ε_r and gives an upper limit of the film's loss factor $\tan\delta$. Resonators with different AlN films were fabricated and characterized. To prove the validity of the gained results a high-pass filter was designed, fabricated and characterized.

3.5.1 Theoretical Background

For ultra-thin films, the characterization of the ε_r is challenging, which is why a special design had to be developed to allow the measurement of all parameters that are needed for the calculation. The ε_r can be determined out of the capacity C of a plate capacitor with AlN as a dielectric with known spatial dimensions:

$$C = \varepsilon_r \varepsilon_0 \cdot \frac{A}{d}. \tag{3.7}$$

ε_0 is the vacuum permittivity, A is the area of the capacitor and d the thickness of the dielectric. To determine a capacity C free from the parasitic capacities caused by feed lines, the resonance frequency f_0 of a LC-resonator can be used with a known inductance L:

$$f_0 = \frac{1}{2\pi\sqrt{LC}}. \tag{3.8}$$

This makes it necessary to accurately determine the inductance of the used resonator. Typically, the inductance of a wire is considered by its magnetic self-inductance which is caused by energy stored in the magnetic field around the wire. Its only depend logarithmically on the wire diameter and its length dependence is in the range of 1 pH/µm [192]. In thin films, a significant amount of the inductance is not caused by its magnetic inductance, further regarded as geometric inductance L_{geo}, but by the inertia of the electrons traveling in the film. The inductance caused by inertia is further regarded as the kinetic inductance L_{kin}. L_{kin} scales with wire length and inversely with the cross-sectional area. In a normal conducing wire the inertia of electrons is limited due to constant collisions with phonons, decreasing their speed. The impedance dominates L_{kin}. However, in a superconductor, no DC resistance is present and electron-pairs can move freely, and are thereby able to reach a significant inertia. From DC to GHz frequencies the impedance can be dominated by the kinetic inductance of the supercurrent. For a 5 nm thick NbN film, the square inductance $L_\square$ is in the range of 120 pH [193] [147]. By careful design with a width of the inductor in the nano scale, one can make sure that $L_{kin} >> L_{geo}$, so only the kinetic inductance needs to be considered. The kinetic inductance of a superconducting film can be calculated using the London penetration depth λ_L:

$$L_{kin} = \mu_0 \cdot \lambda_L(T_B)^2 \cdot \frac{\ell}{A}, \tag{3.9}$$

where ℓ is the length of the inductor and A is its cross section. The London penetration depth at bath temperature T_B, $\lambda_L(T_B)$, can be calculated using eq. (3.10) [194]:

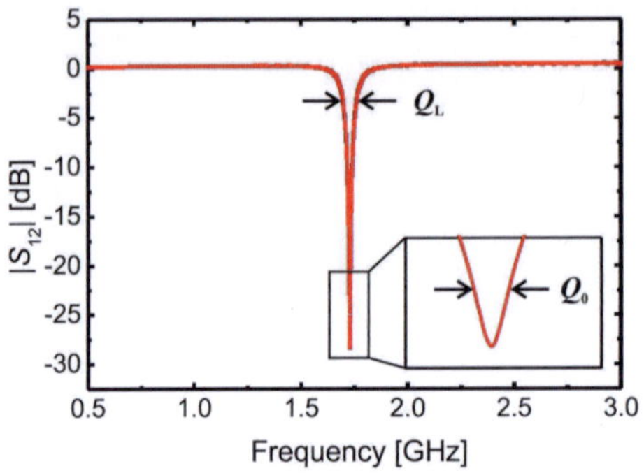

Figure 3.19: Typical resonance dip of a LC-resonator [195]. The positions to determine the loaded Q_L and the intrinsic quality factors Q_0 are marked.

$$\lambda_L(T_B) = \frac{\lambda_L(0)}{\sqrt{1-(T_B/T_C)^4}}, \tag{3.10}$$

where the penetration at $T = 0$, $\lambda_L(0)$, can be calculated using the residual resistivity ρ_n and the critical temperature T_C of the film using eq. (3.11) [194]:

$$\lambda_L(0) = 1.05 \cdot 10^{-5} \cdot \sqrt{\rho_n/T_C}\text{cm}. \tag{3.11}$$

An upper limit of $\tan\delta$ can be determined using the unloaded intrinsic quality factor Q_0 of the resonator (fig. 3.19). The unloaded Q_0 includes losses due to NbN surface resistance Q_ρ, radiative losses Q_{rad} and losses in the dielectric Q_ε:

$$\frac{1}{Q_0} = \frac{1}{Q_\rho} + \frac{1}{Q_{rad}} + \frac{1}{Q_\varepsilon}. \tag{3.12}$$

Q_{rad} can be neglected for a carefully matched resonator design. For a short inductor and a wide conductor compared to the overall size of the resonator, Q_ρ should be small compared to Q_ε. In case of a small Q_ρ, one can assume that $Q_0 = Q_\varepsilon = \tan\delta$ [195]. Q_0 is extracted out of the frequency response of the resonator (a typical resonance dip is shown in fig. 3.19). The Q_0 is calculated out of the bandwidth Δf_{Q_0} of the resonance dip and the resonance frequency f_0:

$$Q_0 = \frac{f_0}{\Delta f_{Q_0}}. \tag{3.13}$$

The attenuation level to extract the Q_0 was calculated using [196]:

$$S_{12,Q_0} = \sqrt{\frac{2 \cdot S^2_{12,f_0}}{1 + S_{12,f_0}}}, \tag{3.14}$$

where S_{12,f_0} is the depth of the resonance dip.
In summary, ε_r can be determined by using a LC-resonator with AlN as a dielectric and measuring the resonance frequency f_0 in combination with measuring T_C and ρ_n of the NbN film. $\tan\delta$ can be determined out of the unloaded quality factor of the resonance dip if the surface resistance of the used NbN film is small enough. A chip design, that allows the measurement of all these parameters had to be developed.

3.5.2 Resonator Design

To determine ε_r and $\tan\delta$, a chip design was developed utilizing 5 superconducting LC resonators embedded in a microstrip transmission line along with two bridges to determine T_C and ρ_n and measurement stripes to determine the thickness of the individual layers (fig. 3.21). A dual-side polished R-plane sapphire of a thickness of 350 µm was chosen as a substrate since it is an excellent low loss microwave substrate with a $\tan\delta < 10^{-5}$ [197] at a temperature of 4.2 K and a well-known substrate for the growth of high-quality NbN films. As a ground plane for the microstrip transmission line, a gold conductive layer ($d > 235$ nm) was deposited onto the backside of the substrate. To demonstrate the feasibility of making an integrated microwave circuit along with a SNSPD from the same film, the thickness of the microstrip transmission line and resonators was chosen to be 5 nm. The geometry of the microstrip transmission line was optimized for the used substrate thickness and for a kinetic sheet inductance of the $L_\square = 120$ pH using the Sonnet EM simulation software [198] to an impedance of 50 Ω. The optimal strip width for an impedance of 50 Ω was found to be 465 µm.

The resonators were designed as lumped-element LC circuits [195]. To reduce the number of required fabrication steps, which improves the yield, a two capacitor design was used (fig. 3.20a). The design allows the fabrication of the AlN dielectric and the top electrode in situ with each other. It consists of two thin-film capacitors with identical capacities (C_1 & C_2) and two inductors (L_1 & L_2). L_1, the microstrip transmission line and the lower electrode of the thin-film capacitors are structured out of a 5 nm thick NbN film. L_2 and the top electrode of C_1 & C_2 are made from a 200 nm thick NbN film. L_1 has a length of 500 µm and a width of 25 µm resulting in a total inductance of $20L_\square$. L_2 bridges the gap between both capacitors and has a length of 50 µm and a width corresponding to the width of capacitors. Due to the

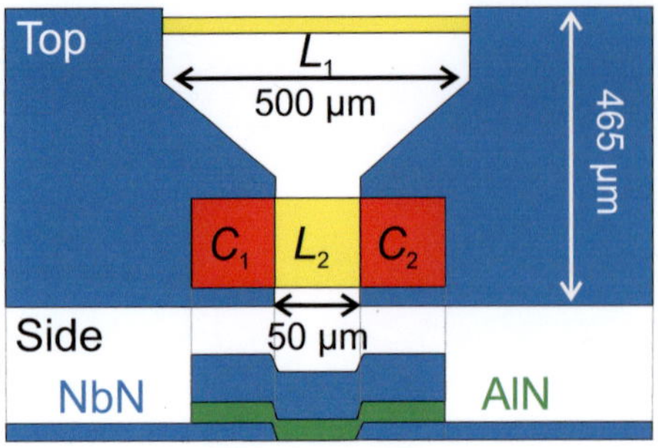

(a) Resonator design (not to scale)

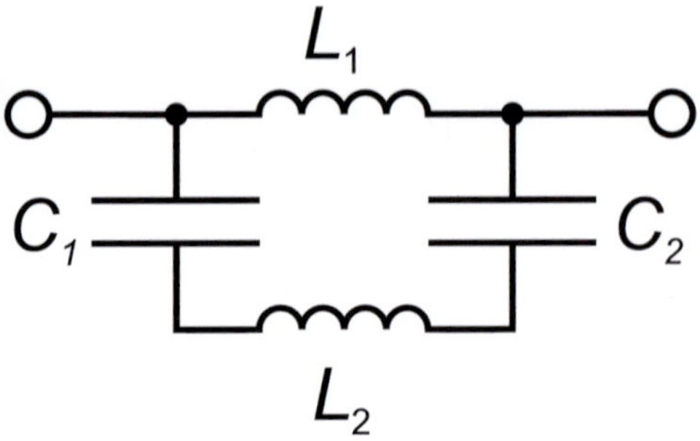

(b) Electrical equivalent circuit

Figure 3.20: Design of a single LC resonator (a) and its equivalent electrical circuit (b). (a) Top view of the embedded LC-resonator with a side view of the layer structure for capacitors. In the top view the microstrip transmission line is shown in blue; the nanowires acting as inductor are depicted in yellow; the capacitors are shown in red. In the side view the NbN electrodes are shown in blue whilst the AlN dielectric is colorized green.

significant lager thickness, the inductance of L_2 is at least 80 times smaller than the inductance of L_1 and can be neglected. As C_1 and C_2 are in series, the capacity of the resonators simplify to $\frac{1}{C} = \frac{1}{C_1} + \frac{1}{C_2}$. In addition, since the gap is 50 µm wide, an occurring parasitic capacity of the resonator can be neglected since it is small in comparison to C. The resonance frequency of the presented resonator design is:

$$f_0 = \frac{1}{\pi \cdot \sqrt{2L_1C}}. \tag{3.15}$$

To achieve an equal spacing of the resonance frequencies, the area of the capacitors was used to adjust f_0. The areas were chosen to 50 µm × 50 µm, 25 µm × 50 µm, 25 µm × 25 µm, 25 µm × 15 µm and 25 µm × 10 µm. This results in a resonance frequency of the individual resonators compared to the resonator with the lowest resonance frequency $f_{0,1}$ of $f_{0,2} = 1.4 \cdot f_{0,1}$, $f_{0,3} = 2 \cdot f_{0,1}$, $f_{0,4} = 2.5 \cdot f_{0,1}$ and $f_{0,5} = 3.1 \cdot f_{0,1}$, respectively.

The full chip design is shown in fig. 3.21. The microstrip transmission line with integrated resonators shown in (a) is connected during measurement to port P1 and port P2. For the calculation of L_{kin}, the ρ_n of the 5 nm NbN film was determined out of the two bridges (b) which contain contacts for a four-point measurement scheme. The longer bridge has a width of 50 µm and a length of 2500 µm and the shorter bridge was chosen to have a width of 25 µm and a length of 500 µm. Both bridges were designed to have width of 500 nm which is significantly larger than width variations of the optical lithography and to have a length of at least 20 times their width, to lower the influence of small defects in the bridge on the measured ρ_n. Several stripes (c) were placed into the design to accurately determine the thickness of individual layers.

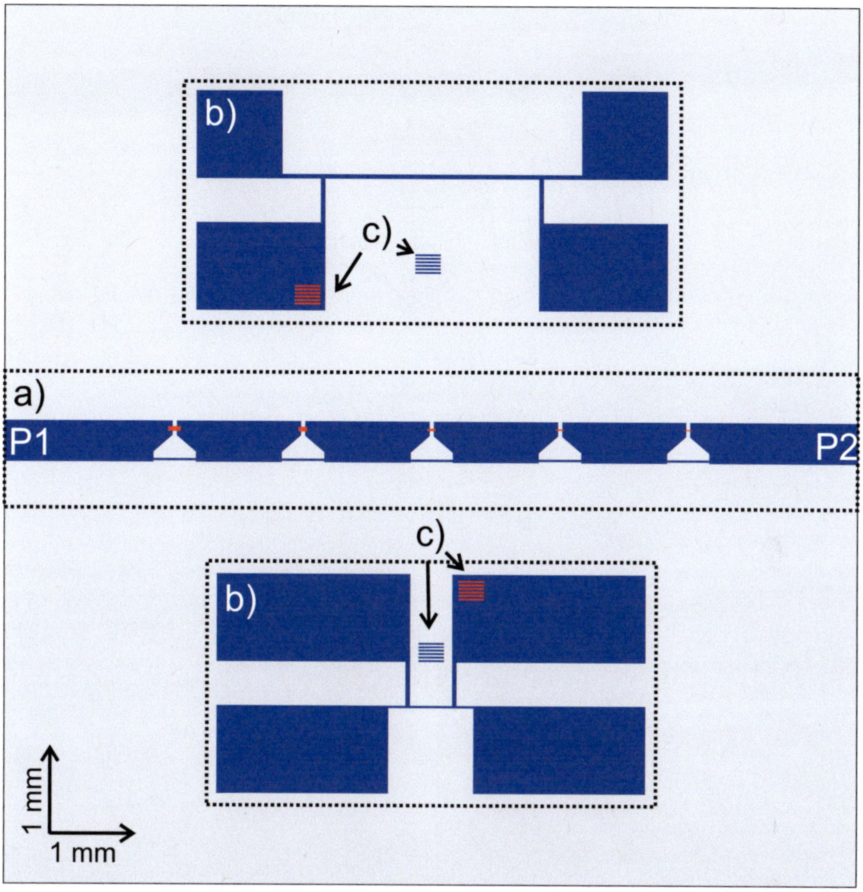

Figure 3.21: Design of the resonator chip. Five resonators are embedded in a microstrip transmission line. The bottom NbN film is shown in blue, the top film in red. a) Microstrip transmission line with 5 embedded LC-resonators; the two ports are indicated with P1 and P2. b) Four-point measurement bridges to determine ρ_n. c) Stripes to accurately determine the thickness of each layer.

3.5.3 Resonator Fabrication

Samples were fabricated with AlN thicknesses from 6.4 to 25 nm, at two different temperatures and deposition rates. The fabrication of a resonator chip was performed in three deposition and two patterning steps (fig. 3.22). In the first deposition step, a 10 nm thick Nb adhesion layer was deposited using sputter deposition at a rate of 0.33 nm/s followed by the deposition of a more than 235 nm thick Au ground layer in situ at a rate of 1.3 nm/s. The chip was then flipped and a 6 nm thick NbN film was deposited. The NbN deposition was done at a temperature of 850 °C in an Ar/N_2 atmosphere at a rate of 0.08 nm/s. The microstrip transmission line, test structures and alignment marks were patterned using photolithography and reactive ion etching in an O_2/SF_6 atmosphere. In the third deposition step, the AlN dielectric layer was deposited followed by an in situ deposition of a 160 nm thick NbN layer

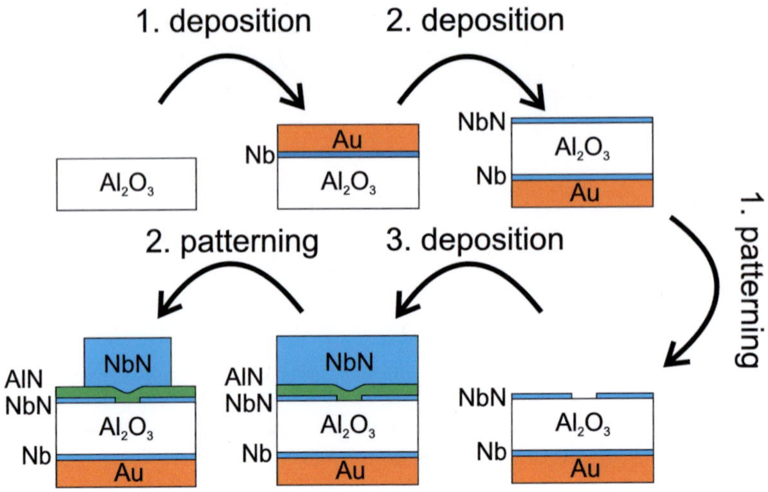

Figure 3.22: Fabrication process of resonators

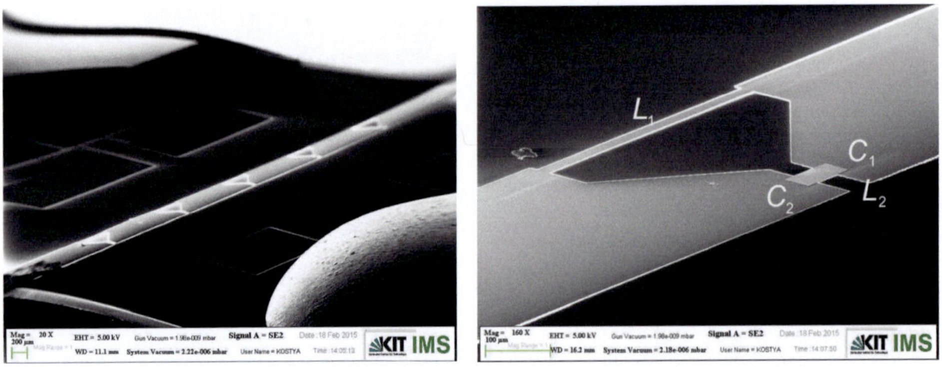

(a) Full chip

(b) Single resonator

Figure 3.23: Tilted SEM images of a full resonator chip and and a single resonator.

serving as a top electrode for the thin film capacitors. The AlN was deposited at different thicknesses at 36 °C and 600 °C and at deposition rates of 0.09 nm/s and 0.04 nm/s. The top NbN layer was patterned using two photolithography and RIE steps: first, windows were etched into the top electrode to reveal alignment marks for a second lithography step. In the second lithography, the capacitors were defined. The AlN layer hereby acts as an etch stopping layer for the RIE step, protecting the micro-strip transmission line from being etched. Finally, the AlN film, which is not protected by the top electrode, is stripped by wet etching in a MIF 300-47 developer. A SEM image of the fully fabricated device can be seen in fig. 3.23a and a SEM image of a resonator is shown in fig. 3.23b.

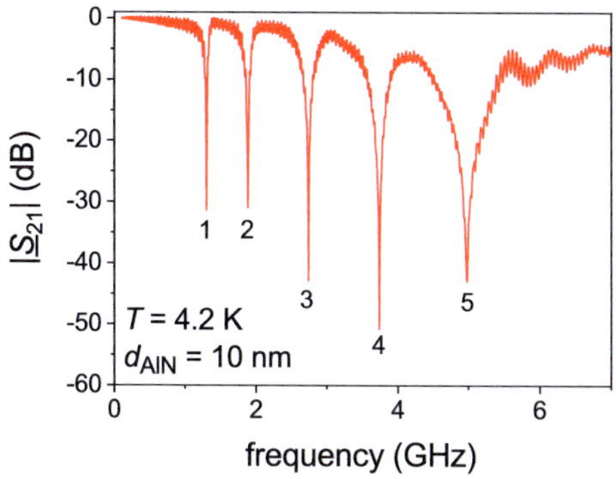

Figure 3.24: Measured transmission for a 10 nm thick AlN dielectric layer deposited at a temperature of 36 °C at a sputter rate of 0.09 nm/s.

3.5.4 Resonator Characterization

The film resistivity and critical temperature was determined by using on-chip bridges for DC characterization in a four-point measurement. The NbN-films have a ρ_0 of 260 ± 10 µΩ cm and a critical temperature of 12.4 ± 0.2 K. From the resistivity the kinetic inductance calculates to 100 ± 5 pH per square using eq. (3.9), eq. (3.10) and eq. (3.11)).

The resonance frequencies of the resonators were characterized for all samples at a temperature of 4.2 K by measuring the transmission characteristics ($|\underline{S}_{21}|$) using a network analyzer at a power of −27 dBm in the frequency range from 0.1 to 7 GHz. The resulting $|\underline{S}_{21}|$ is shown for a sample with an AlN thickness of 10 nm deposited at a deposition rate of 0.09 nm/s at a temperature of 36 °C in fig. 3.24. All five resonance dips are clearly visible. At 6 GHz some calibration artifacts are visible: the calibration was done at room temperature and not at measurement temperature. The resonance frequencies are read out at the minimum of the resonance dips. The dips can be assigned to the corresponding resonators by assigning the lowest resonance (1) to the resonator with the largest capacitors and so on. For the sample shown, the extracted resonance frequencies are in ascending order: $f_{0,1}$ = 1.30 GHz, $f_{0,2}$ = 1.88 GHz, $f_{0,3}$ = 2.73 GHz, $f_{0,4}$ = 3.73 GHz and $f_{0,5}$ = 4.98 GHz. In relation to $f_{0,1}$, the resonance frequencies are $f_{0,2} = 1.45 \cdot f_{0,1}$, $f_{0,3} = 2.1 \cdot f_{0,1}$, $f_{0,4} = 2.87 \cdot f_{0,1}$ and $f_{0,5} = 3.83 \cdot f_{0,1}$. In comparison to the designed frequency spacing, the resonances shift toward higher frequencies for resonators at higher resonance frequencies. This is caused by a not fully symmetric placement of the top electrode, which increases the area of one capacitor and decreases the area of the other capacitor for a pair of corresponding capacitors. Hence, the effective capacity is decreased in comparison to a symmetric capacitor pair. This issue was solved by a measuring the area of the capacitors using an optical microscope and calculating the effective area of the "combined" capacitor. The effective capacity for each resonator was extracted out of the resonance frequency using eq. (3.15). The thickness of the AlN dielectric was measured with a profilometer at the thickness measurement stripes (fig. 3.21c). The relative permittivity ε_r was calculated out of the thickness, the area of capacitors and the measured capacity.

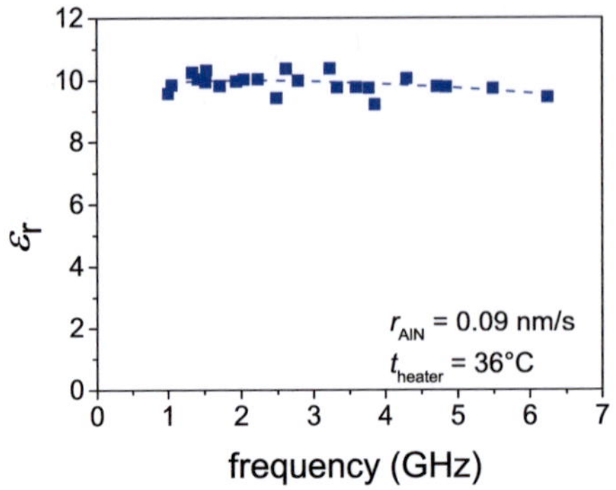

Figure 3.25: Frequency dependence of the ε_r. The data stems from 5 samples with 5 resonators each and a thickness of the AlN dielectric from 8 to 25 nm. The dashed line is to guide the eye.

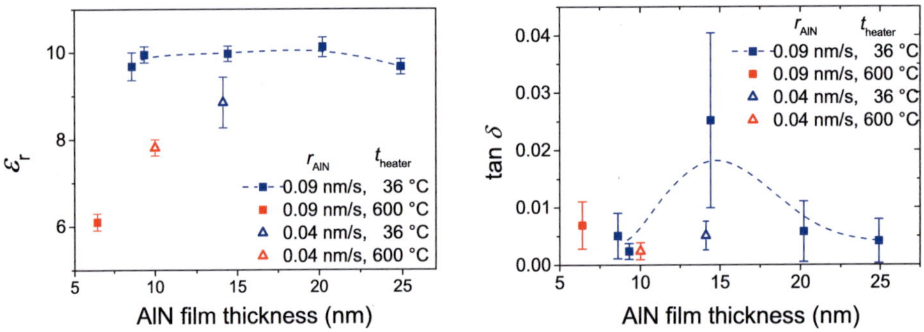

(a) ε_r (b) $\tan\delta$

Figure 3.26: Influence of the AlN thickness on relative permittivity (a) and dissipation factor (b). Each point corresponds to one fabricated sample and is an average for the 5 resonators. The error bars indicate the spread of extracted values of ε_r and $\tan\delta$ for the investigated resonators. Films fabricated at a low temperature are indicated with blue and at high temperature with red symbols. Films indicated with open symbols were deposited at a rate of 0.04 nm/s and filled symbols with 0.09 nm/s.

The frequency dependence of ε_r for all characterized resonators with an AlN dielectric deposited at a deposition rate of 0.09 nm/s at a temperature of 36 °C is shown in fig. 3.25. The figure includes the ε_r of all 5 resonators for 5 samples with a thickness of the AlN dielectric from 8 to 25 nm (see fig. 3.26). The ε_r remains almost constant at 9.9 ± 0.3 with a slight decrease towards higher frequencies.

The dependence of ε_r and $\tan\delta$ on the thickness of the AlN and on different deposition conditions is shown in fig. 3.26. Each data point is averaged for the 5-resonators of each sample. No thickness dependence of the ε_r for samples deposited with identical deposition conditions can be seen. The deposition conditions of AlN significantly influence the dielectric properties of the AlN. Films deposited at a low temperature and a deposition rate of 0.09 nm/s have a ε_r of 9.9 ± 0.3. This value is in good agreement with literature values of ε_r for AlN grown in a wurtzite lattice: ε_r = 10.31 [199], ε_r = 10.5 [200]. For a deposition rate of

0.04 nm/s, the ε_r decreases to 8.8. An increase in the deposition temperature to 600 °C decreases the ε_r to 7.8. A $\varepsilon_r = 6.1$ was measured for $r_{\mathrm{AlN}} = 0.09$ nm/s at a deposition temperature of 600 °C. However, the very small thickness of the AlN dielectric may have influenced the result in this case, since the dielectric has to bridge a 6 nm thick step from the lower NbN electrode to the substrate in between both capacitors (fig. 3.20a). With a thickness of 6.4 nm for the AlN dielectric, the excess thickness is only 0.4 nm which may be to thin too reliably isolate. The strong dependence on deposition conditions indicates a strong influence of the AlN crystalline structure on ε_r. The dissipation factor $\tan\delta$ is in the range of 0.005 for most films. However, there is a strong deviation between samples. In addition, the spread of values for $\tan\delta$ for the different resonators on the same sample is significant. This value is in good agreement with the $\tan\delta = 0.003$ for sputtered films found by [200]. Samples deposited at a lower deposition rate are on the lower end of the measured losses for all films. A strong outlier is the sample with an AlN thickness of 14.4 nm. In addition, this sample has a strong spread for measured values of the individual resonators. This indicates fabrication issues with this sample. One has to note that the design was not optimized for the characterization of losses. The neglected losses due to edge effects or losses in the NbN top electrode may have influenced the results. Since parasitic losses were not excluded, the measured loss values should not be seen as actual loss values for the deposited AlN, but as an upper boarder: the losses in the AlN itself are expected to be lower.

3.5.5 High-Pass Filter

To proof the compatibility of the fabricated AlN as a dielectric for a capacitor in the high-pass path of an integrated bias tee, a high-pass filter based on a coplanar layout was designed, simulated and characterized. Since the pulse of a SNSPD has a quite broad frequency spectrum with a maximal power density between 50 MHz and 1 GHz [191], the high-pass filter should have a cut off frequency as low as possible and be of compact size. Furthermore, the design should be reliable to fabricate. The NbN film was chosen to exhibit a thickness of 6 nm to enable a later detector integration along with the high-pass filter. The coplanar transmission line was designed to be matched to an impedance of 50 Ω as used for the readout. To obtain an impedance of 50 Ω the inner conductor of the coplanar transmission line has a width of 350 µm and a gap to the ground plane of 155 nm. Towards the sides of the chip a linear taper increases the width of the inner conductor to 500 µm and the width of the gap to 195 µm to maintain the impedance over the full length and to allow the use of standard SMA connectors. The size of the capacitor for the high-pass filter on-chip was chosen to a maximum size of 350 µm × 500 µm, which corresponds to the size of contact pads for the SNSPDs on GaAs used in later chapters. The capacitor is placed above a gap of 50 µm in the inner conductor of the coplanar transmission line in a two capacitor design (fig. 3.27a). The width of the capacitors corresponds to the width of the inner conductor and has an area of 350 µm × 235 µm. The thickness of the AlN dielectric was chosen to be 20 nm to ensure good isolation properties. The top electrode has a thickness of 200 µm and is made of NbN. The sample was fabricated using the same steps as for the resonator fabrication with an adapted design. The design was simulated using the Sonnet EM simulation software with a $\varepsilon_r = 9.8$ and a $\tan\delta = 0.003$ for the AlN dielectric and a sheet kinetic inductance of 100 ± 5 pH for the

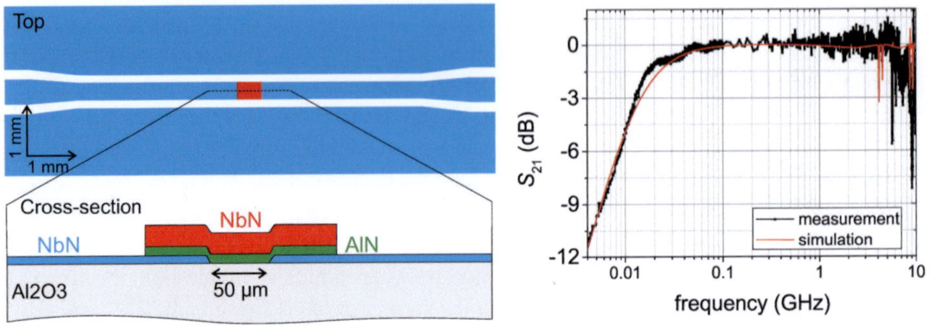

(a) High-pass filter design (b) Transmission characteristics

Figure 3.27: (a) Design of the high-pass filter from the top (top) and the cross-section of the capacitors from the side (bottom). The NbN base layer is shown in blue, the AlN dielectric is shown in green and the top electrode is shown in red. (b) Comparison of the measured and the simulated frequency dependence of the high-pass filter.

NbN film. The top electrode was simulated as lossless metal as the kinetic inductance of the top layer is significantly lower than the inductance of the bottom layer due to its significant larger thickness. Out of the simulation results shown in red in fig. 3.27b, a design cutoff frequency of 14.6 MHz for the high-pass filter can be extracted. At ≈4.1 GHz a dip is visible that limits the effective 3 dB bandwidth of the high-pass filter to ≈4 GHz which would be sufficient for the use for a SNSPD readout.

The transmission of the high-pass filter was characterized at 4.2 K in a range spanning from 3 MHz to 10 GHz. The measured high-pass filter characteristic fits well with the expected characteristic from the simulation. The cutoff frequency is measured to be 12.9 MHz. The measured 3 dB bandwidth is 5.6 GHz. The good agreement of the simulation with the measured results show that the AlN film can be fabricated with deterministic properties and can be used as a dielectric for on-chip high-frequency integrated circuits.

3.6 Summary

In this chapter, the deposition of thin NbN films with AlN buffer layers was investigated. The optimal deposition temperature was found to be limited by the annealing temperature used for integrated QDs to prevent QD degradation and not by a decreasing NbN film quality. By a careful optimization of deposition parameters a T_C of 12.1 K, a ΔT_C of 1.1 K and a *RRR* of 0.9 was achieved for a 5.5 nm thick NbN film on GaAs, in combination with a 12 nm thick AlN buffer layer. This is an improvement of T_C by ≈3 K in comparison to NbN films deposited directly on GaAs during this thesis and in comparison to previously reported T_C values for thin NbN films on GaAs (8.2 K [81]; 8.7 K [80]; 9.4 K [189] all values were adapted to the 0.1 % ρ_n criterion used for the determination of T_C during this thesis). Optimal parameters for the deposition of the AlN buffer layer were developed in order to improve the NbN film quality. Already a 2.5 nm thin AlN buffer layer significantly improves the quality of the NbN film. The optimal thickness for the AlN buffer layer was found to be 12 nm. At this thickness, AlN acts as an efficient diffusion barrier for oxides and arsenic. For larger thicknesses, an increase in the surface roughness of the AlN decreased the quality of the NbN film. Nevertheless, it was found that a NbN film grown on a 100 nm thick AlN film still has a T_C = 10.9 K, which is sufficient for a SNSPD operating at 4.2 K. This would allow the use of a NbN SNSPD on a sputtered AlN photonic integrated circuit [201] deposited with the optimal deposition parameters developed in this chapter. The investigation of the ε_r and $\tan\delta$ suggests a wurtzite growth of sputtered AlN films. At room temperature sputtered AlN films exhibit a $\varepsilon_r = 9.9 \pm 0.3$ and $\tan\delta < 0.005$. The usability of the deposited AlN as a dielectric for on-chip high frequency integrated circuits was demonstrated by the on-chip integration of a high pass filter with a 3 dB bandwidth of 5.6 GHz.

4 NbN-Based SNSPD with AlN Buffer Layer for GaAs Waveguide Integration

For fully integrated GaAs photonic circuits, high-quality detectors are needed with an high DE for the wavelength of integrated QDs. In the previous chapter, the quality of thin NbN films grown on GaAs was significantly improved towards a higher T_C, I_C and RRR by the use of an AlN buffer layer. To examine if the modified film properties are beneficial to improve the properties of a SNSPD, NbN SNSPDs on GaAs were fabricated out of NbN films with and without AlN buffer layer and investigated regarding their detection properties. This investigation is necessary since a maximization of superconducting characteristics does not necessarily correspond to better detection properties. A SNSPD design was developed suitable for the direct integration into a GaAs waveguide-based photonic chip. Results presented in this chapter are published in part in IEEE Transactions on Applied Superconductivity under the title *"AlN-Buffered Superconducting NbN Nanowire Single-Photon Detector on GaAs"* [SIS17]. In this chapter, the detector design is introduced and the fabrication and pattering is described. The used fiber-coupled optical measurement setup is introduced. SNSPDs are characterized regarding their superconducting and optical properties and their usability for photonic integration is evaluated.

4.1 Detector Design

The detector was designed so that it can be integrated into an optical waveguide (WG) after a WG taper (see section 2.3.8 in chapter 2) that increases the width of the WG to 10 µm.

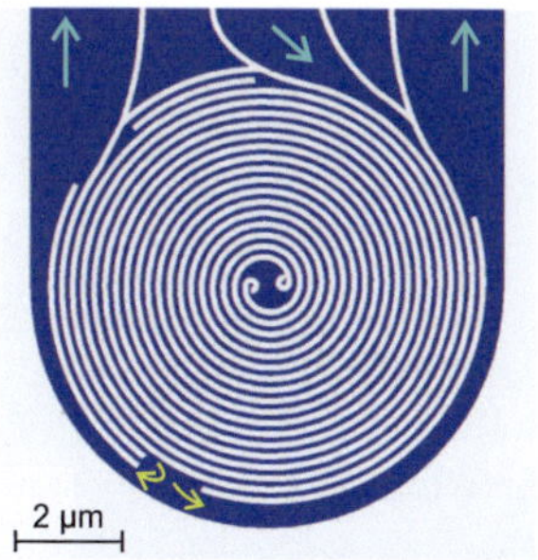

Figure 4.1: Design of the studied double-spiral SNSPD. The NbN is shown in blue and the substrate in gray. The current flow in and out of the detector is indicated by light blue arrows. Yellow arrows indicate the current splitting into the ground plane of the RF coplanar WG at the end of the nanowire.

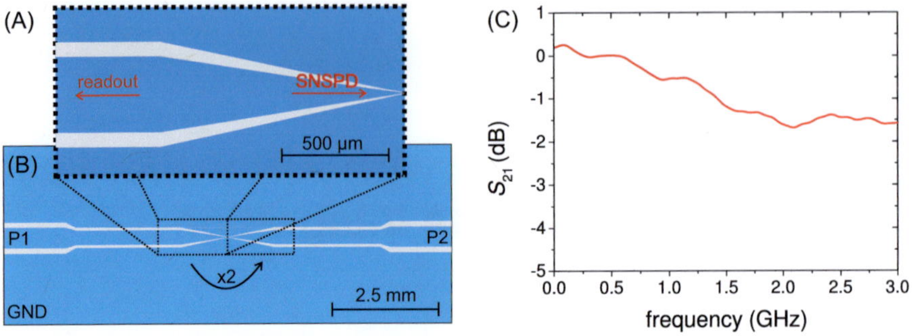

Figure 4.2: (a) RF coplanar waveguide with taper to readout and bias the SNSPD, the substrate is shown in gray and the NbN in blue. (b) Coplanar test design to test the RF properties of the taper. The ports are indicated with P1 and P2. (c) Transmission loss measured for the test design shown in (b). The data was smoothed using FFT filtering to remove standing waves in the feed lines.

To reduce current crowding and enhance the critical current and the corresponding detection efficiency, a double-spiral design was chosen [161]. The spiral design is insensitive to linear polarization, hence the TE as well as the TM mode on WG have the same absorption efficiency. The spiral consists of two, archimedian spirals running in parallel that are connected in the middle (fig. 4.1). The detector was designed to a total diameter of 7.6 µm to fill out the available space on the taper. In the middle of the detector, two 3/4 circles increase the radius of the inner bend to $r = 150\,\mathrm{nm}$ in order to reduce current crowding. At the end of the nanowire, the current is split into the ground planes of the RF-readout line. To the side, additional spiral segments are added to reduce the influence of the proximity effect during electron beam lithography on the actual spiral. With a width of 120 nm for the nanowire and a gap of 80 nm between wires, the total length of the detector is 218 nm. The thickness was designed to 5 nm.

The detector is connected to a NbN coplanar radio-frequency waveguide (RF-WG) with a thickness of the SNSPD. The RF-WG consists of a linear taper and a short RF-WG section to electrically contact the detector for biasing and readout of the detector pulse (fig. 4.2a). The coplanar taper was designed to be matched to the impedance of the readout chain of 50 Ω. The taper starts at the detector with a width of the inner conductor of 2 µm and a gap of 1.3 µm with a wave impedance of 70 Ω and is linearly broadened to a width of 340 µm with a gap of 66 µm with a wave impedance of 50 Ω. The RF-WG and taper were characterized in a transmission measurement. To obtain an upper limit of the attenuation of the on-chip RF-WG, a design was developed to characterize the taper in a transmission measurement (fig. 4.2b). The designed taper was placed two times in the middle of the chip, with the small width sides facing each other. Behind taper an additional coplanar RF-WG was placed to allow the electrical contact of the inner tapers in a 10 mm × 10 mm sized sample holder. Towards the ends of the RF-WG, additional tapers increase the width of the inner conductor to a width of 500 µm whilst maintaining an impedance of 50 Ω. The width of 500 µm of the inner conductor at the end allows the direct placement of SMA-connectors on the coplanar line for electrical contact. The characterization was done using a network analyzer in a frequency range from 50 MHz to 3 GHz (fig. 4.2c) at a temperature of 4.2 K. Since the measured transmission ($|\underline{S}_{21}|$) contains

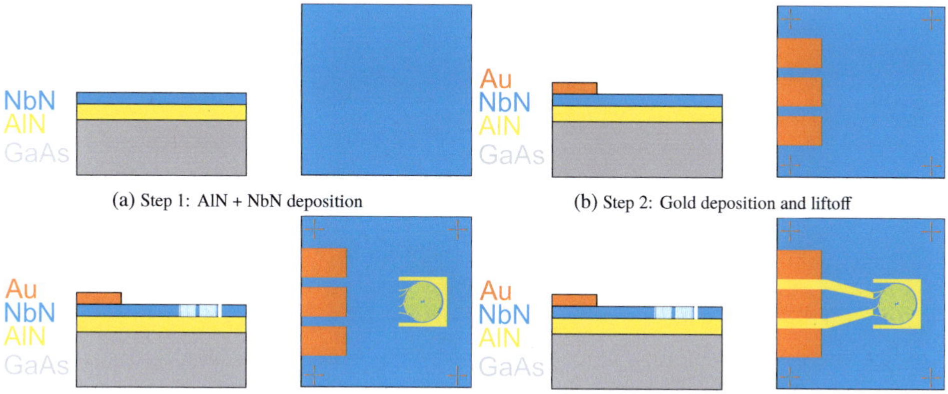

(a) Step 1: AlN + NbN deposition

(b) Step 2: Gold deposition and liftoff

(c) Step 3: Detector etching

(d) Step 4: Gap etching

Figure 4.3: Fabrication steps for a spiral SNSPD on GaAs with AlN buffer layer.

two tapers and a longer coplanar contact line, the upper limit for the attenuation of a single taper can be estimated by $|\underline{S}_{21}|/2$. The upper limit for the transmission loss in a single taper is ≈0.05 dB from 50 to 500 MHz, which increases to ≈0.8 dB in the range from 1.5 to 3 GHz.

4.2 SNSPD Fabrication

The detector fabrication and pattering was done in a four step process. In a first step, AlN and NbN is deposited (fig. 4.3a). 10 nm of AlN is deposited onto a bare GaAs(100)-substrate using the previously developed sputter deposition process (chapter 3). The NbN film is sputtered to a thickness of 5 nm on top of the buffered substrate and on top of a second bare GaAs(100) substrate in the same deposition run to ensure identical deposition parameters for both substrates. In the second step, alignment marks and contact pads are fabricated (fig. 4.3b). The alignment marks consist of gold crosses and were used to accurately adjust the position and rotation of all consecutive lithography steps to each other. Gold was used since it has a high electron contrast to NbN and GaAs and can be used for automatic alignment by the electron beam lithography (EBL) system. For the electron beam lithography a 160 nm thick PMMA 950K electron beam resist was used. PMMA stands for poly(methyl methacrylate) and is a positive tone electron beam resist. 950K hereby stands for the molar mass of the polymer chains in the resist. After development of the resist, a 10 nm thick Nb adhesion layer followed by 30 nm of gold is sputtered. Afterward, the gold and Nb is lifted in unwanted areas by removal of the resist using acetone. The detector is patterned in a third process step using EBL with a 70 nm thick PMMA 950K resist. After development and prior to etching, the resist was hardened by the exposure of a 8 µm × 8 µm square on top of the detector using EBL. This post exposure process ensures a smoother resist profile and reduces the edge roughness, leading to a smaller number of defects in the final device. The detector is structured out of the NbN film using reactive ion etching (RIE) in a SF_6 and O_2 atmosphere (fig. 4.3c). AlN and GaAs are not attacked by SF_6 and O_2 in a reactive process and act as natural stopper layers and

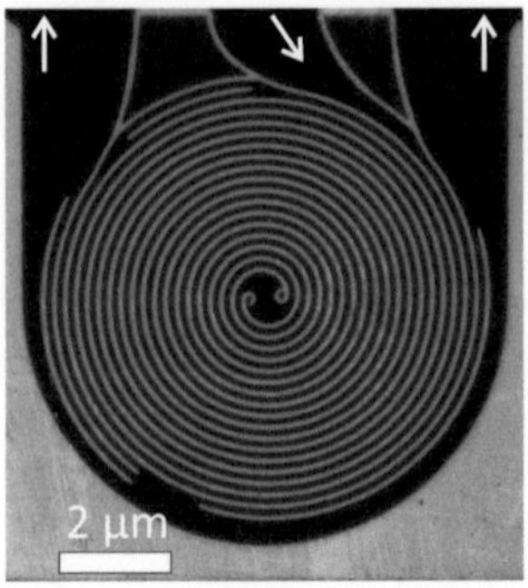

Figure 4.4: SEM image of a fully patterned double spiral SNSPD. The AlN op top of the GaAs substrate is visible in light gray and the NbN is visible in dark grey. Arrows indicate the direction of the current flow.

consequently, underlying WG-structures should not be influenced by the patterning process. In a final process step, the coplanar RF-WG is structured using electron beam lithography and RIE. A SEM image of a detector after fabrication can be seen in fig. 4.4. The fabricated SNSPDs have a thicknesses of 5 nm, nanowire widths of 120 nm, a gap sizes between wires of 80 nm and lengths of 218 µm.

4.3 Experimental Setup

The SNSPD properties were characterized in two separate setups. The transport properties were investigated using a simple four probe dipstick setup in a liquid helium dewar. The temperature is measured using a Si-diode temperature sensor with an accuracy of ±0.5 K placed at the detector position. The helium gas in the Dewar naturally develops a temperature gradient, from room temperature down to the 4.2 K of the liquid helium. The temperature during measurement can be adjusted by changing the sample position in the gas. This allows the characterization of the detectors resistivity from 4.2 to 300 K. The resistance was measured as a function of the temperature during cooling down. Current-voltage dependencies were recorded at a temperature of 4.2 K.

The optical properties were studied in a vacuum-dipstick cryostat with fiber optical access. The principle of the full setup can be seen in fig. 4.5. A detailed description of the cryostat can be found in chapter 3 of [191]. The detector is glued to a copper sample holder with conducing silver adhesive. The sample holder has two rings of copper-beryllium springs that touch the thermal shield to ensure a good thermal connection. The dipstick is placed into a liquid helium Dewar to keep the detector at 4.2 K during optical characterization. Inside the dipstick, the detector connects electrically to a low temperature bias tee made from distributed elements. The detector is biased through the DC side and the detector output pulses are split to the RF side of the bias tee. The DC side is connected to a room-temperature low-noise battery driven bias source. The RF side is connected to a semi-rigid RF coaxial cable with an impedance of 50 Ω. The pulses are amplified by three cascaded room temperature amplifiers with a total amplification of 57 dB, a bandwidth of 2.9 GHz and low-frequency cutoff of 50 MHz.

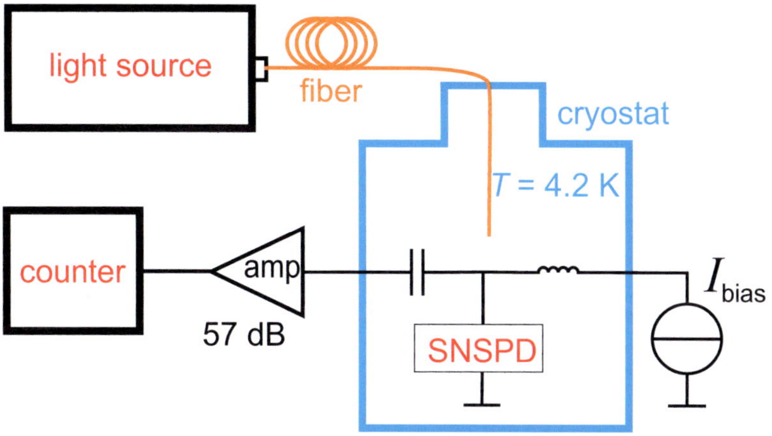

Figure 4.5: Principle diagram of the fiber-coupled optical setup for SNSPD characterization.

The first amplifier in the chain is a Mini Circuits ZX60-P103LN+ with a noise figure of 0.6, followed by two Mini Circuit ZFL-2500VH amplifiers. The optical signal travels through a multi-mode fiber with a core diameter of 105 µm, which ends approximately 3.5 mm above the detector and illuminates the detector perpendicular to the substrate plane from the top. Due to the core diameter and the distance of the fiber end to the detector, the diameter of the light spot is approximately 1.5 mm at the detector plane. The position of the spot can be aligned to the detector with the help of two micrometer screws. For detector characterization, the spot is adjusted for the maximal count rate on the SNSPD. Three different light sources were used for optical characterization. The first light source is a LED light source with a wavelength of 385 nm. The second is a halogen light source in combination with a monochromator, which allows for spectral characterization in the range from 400 to 1700 nm. The spectral bandwidth of the used monochromator settings is 10 nm from 400 to 1050 nm and 20 nm from 1050 to 1700 nm. The third available light source is an erbium-fiber-based fs-laser (C-Fiber, Menlo Systems) with a wavelength of $\lambda = 1550$ nm, a repetition rate of 100 MHz and a pulse length below 150 fs, which is used to investigate the timing jitter of the detectors. Prior to detector characterization, the system is calibrated at room temperature. The input power of all used photon sources is measured with a calibrated photo diode at the detector position. The intensity distribution at the detector position is recorded using a CCD chip. From the known intensity distribution and the known power, the photon flux on the detector area is calculated. Detector pulses are counted by a Stanford Research Systems SR620 counting electronics with an internal bandwidth of 300 MHz and can be directly recorded using an Agilent Infiniium 90000 X-Series real time oscilloscope with a bandwidth of 32 GHz and a sample rate of 80 GS/s.

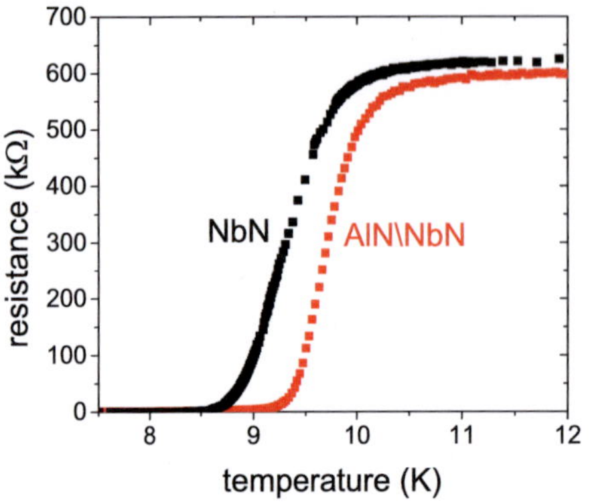

(a) R-T dependence

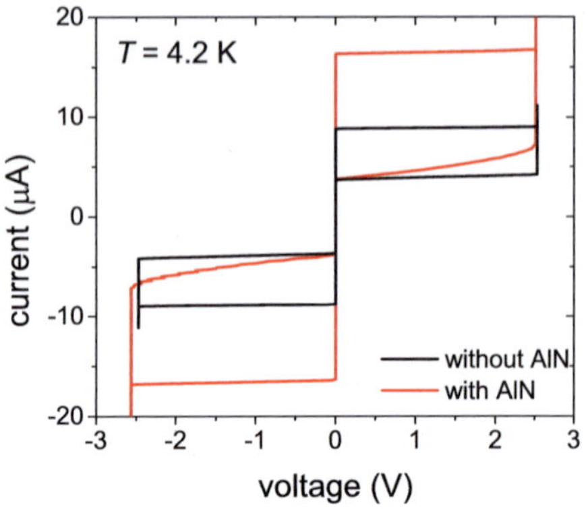

(b) IV dependence

Figure 4.6: Comparison of the superconducting properties for the SNSPD with (red) and without (black) AlN buffer layer. (a) Resistance-temperature dependence; (b) Voltage-current dependence. At voltages above 2.5 V, the used analog-digital converter of the measurement system is saturated and a constant voltage in dependence of the current is visible in the data.

4.4 Electrical and Optical Characterization

SNSPDs on AlN and directly on GaAs were evaluated by characterizing their superconducting and optical properties.

The SNSPDs were patterned out of NbN films with a $T_C = 9\,\text{K}$ ($T_C = 10.3\,\text{K}$ with AlN buffer layer) and a $\rho_n = 278\,\mu\Omega\,\text{cm}$ ($\rho_n = 177\,\mu\Omega\,\text{cm}$). The resistance-temperature dependence of both detectors is shown in fig. 4.6a. With 620 kΩ, the measured resistivity for the SNSPD directly on GaAs is lower than estimated when calculated with the ρ_n of the film (≈1 MΩ), whilst the resistivity of the buffered SNSPD is very close to the estimated value of 590 kΩ measured in comparison to an estimated 641 kΩ. This difference can be explained by a resistive shunting effect of the SNSPD by the substrate, which is omitted by the isolating AlN layer. Out of the superconducting transition, a T_C of 8.7 K can be extracted for the SNSPD on GaAs and of 9.3 K for the buffered SNSPD. The I_C (fig. 4.6b) for SNSPDs is significantly enhanced by the AlN layer. At a temperature of 4.2 K, the I_C is increased from 8.8 μA to 16.5 μA by the presence of the buffer layer with an identical I_r of 3.8 μA. Using I_r in combination with the measured ρ_n of the films, the thermal coupling coefficient B can be estimated (see eq. (3.2) in section 3.2). With $B = 3560\,\text{W}\,\text{K}^4/\text{m}^2$ the thermal coupling is significantly stronger for the NbN film on a bare substrate ($1290\,\text{W}\,\text{K}^4/\text{m}^2$ for the buffered film).

The output pulses of the detectors with and without buffer layer are shown in fig. 4.7 for $I_B = 0.9 I_C$. Due to the higher critical current of the buffered detector, the pulse height is significantly higher at a comparable noise level which results in a higher signal compared to the electrical noise.

The dependence of the detection efficiency (DE) and dark count rate (DCR) on the relative bias current, I_B/I_C is shown in fig. 4.8a. The DCR for the buffered detector is three orders

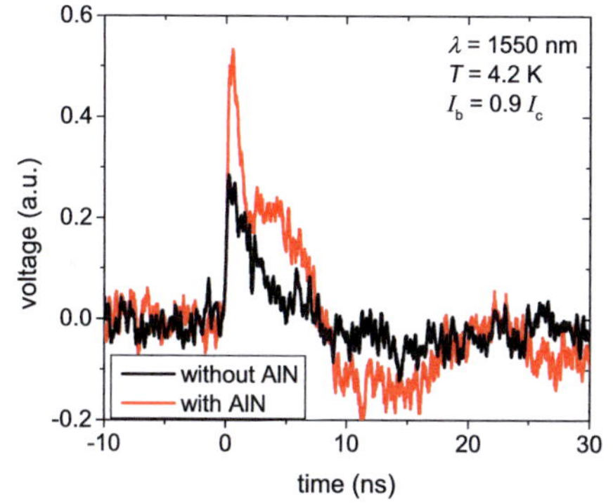

Figure 4.7: Response pulses of the SNSPD with (red) and without (black) AlN buffer layer.

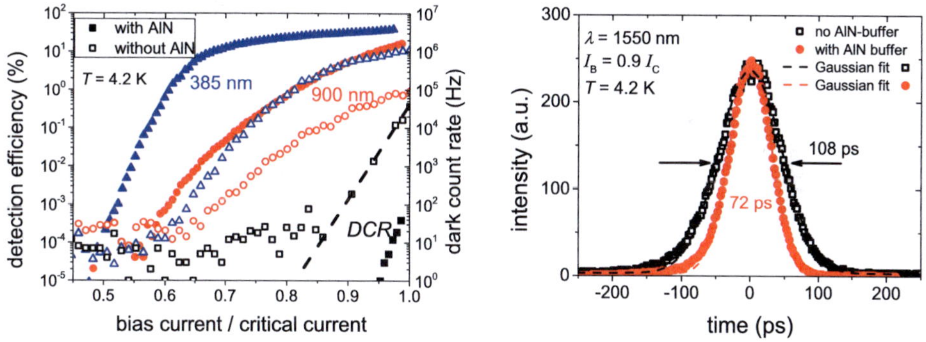

(a) DE and DCR

(b) Jitter

Figure 4.8: Influence of an AlN buffer layer on the DE, DCR (a) and timing jitter (b). The detector without buffer layer is shown in open symbols, the detector with buffer layer in closed symbols. (a) The DE at $\lambda = 385\,\mathrm{nm}$ is shown in blue triangles, at 900 nm in red dots and dark counts are shown in black. The dashed line, is an exponential fit to the DCR of the non-buffered detector above $0.85 I_C$. (b) The timing histogram of the detector without buffer layer is shown in black and the detector with buffer layer in red. The dashed lines are Gaussian fits to the data. ©2017 IEEE [SIS17] of magnitude lower than the DCR for the non-buffered detector. Furthermore, a significant amount of electrical noise can be seen for the non-buffered SNSPD below a bias current of $0.85I_C$. To separate electrical noise from intrinsic dark counts, the DCR at a high bias current above $0.85I_C$ was fitted by an exponential fit. This can be done as intrinsic dark counts have an exponential dependence on the bias current [146] and are several orders of magnitudes higher than dark counts caused by electrical noise at $I_B \mapsto I_C$. The higher electrical noise for the non-buffered SNSPD is caused by a lower trigger level of the pulse counter during measurement. The trigger level was just above electronic noise level since the non-buffered detector has a lower height of the output pulse (fig. 4.7). For $\lambda = 385$ nm the DE for the buffered detector starts to level out at bias currents above $0.75I_C$. It is not saturated and further increases from DE = 19 % at $0.75I_C$ to DE = 31 % at $0.9I_C$. The further increase in DE may be caused by imperfections in the detector. The edge of the nanowire is not perfectly smooth but rather has a roughness of ≈5 nm and, consequently, a local variation of the wire-width. This causes a local change of the current density in the nanowire. Some parts of the nanowire are sufficiently biased, while others are under biased. By increasing the bias current, a larger portion of the nanowire operates in the deterministic detection regime and the effective area of the detector with an optimal efficiency increases and the DE increases further. At a bias current of $0.9I_C$, the DE of the buffered detector is six times higher than for the non-buffered SNSPD on GaAs. At $\lambda = 900$ nm no saturation for both detectors is visible, nevertheless, the DE for the buffered detector is one order of magnitude higher. The increase of the DE can be explained by a better quality of the NbN film due to the buffer layer. At $I_B > 0.97I_C$ the buffered detector starts to latch, which was not observed for the detector without AlN. The latching may be explained by the lower thermal coupling of the NbN to the substrate. The system timing jitter seen in fig. 4.8b was characterized at a $\lambda = 1550$ nm using pulsed-laser light with a pulses length < 150 fs at a bias level of $0.9I_C$. The jitter was determined using the laser trigger as a time reference. The system jitters for both types of detectors follow a Gaussian distribution. The FWHM jitter extracted out of the Gaussian is 72 ps for the AlN buffered and 108 ps for the reference detector. The timing jitter of the SNSPD is reduced by the buffer layer by 36 ps. This can be explained by a lower influence of noise on the jitter of the buffered detector: since the pulses are higher, the relative influence of noise on the jitter is lower [202].

The spectral dependence of the DE from $\lambda = 400$ to 1700 nm at $I_B = 0.9I_C$ is shown in fig. 4.9a. In direct comparison, a higher detection efficiency for the buffered detector is observed. With decreasing photon energy DE also decreases. Between 810 nm and 840 nm an increase in the DE by a factor of two is visible for both detectors. This increase in detection efficiency is connected to the direct bandgap of GaAs of 1.519 eV [203] at the operation temperature of 4.2 K. This corresponds to a wavelength of 817.3 nm which fits well with the observed λ of the observed jump in DE: at shorter λ, the energy of the incoming photons is greater than the bandgap and photons are absorbed by the generation of electron-hole pairs in the semiconductor. For longer wavelengths, the photon energy is too low to excite band transitions and the substrate gets transparent. Consequently, photons pass through the substrate and are reflected at the substrate sample-holder interface. Reflected photons can be absorbed by the detector and the absorption efficiency (ABS) increases, which is observed by the rise of the DE. For short wavelengths, a flattening of the DE can be seen on the buffered detector, indicating that the detector is close to the deterministic detection regime

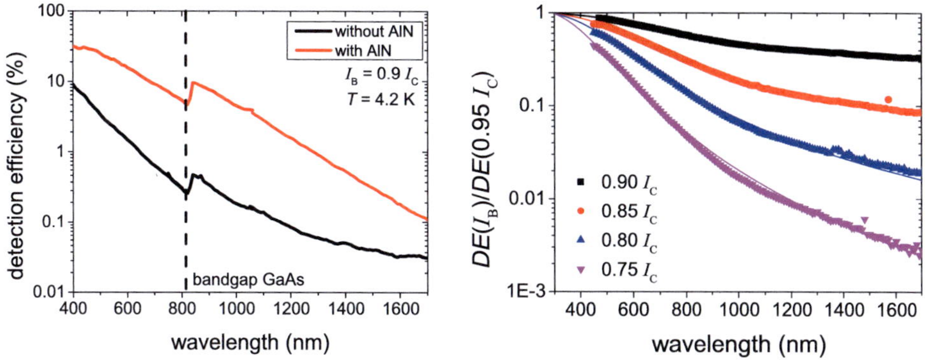

(a) Spectral dependence of the DE for both SNSPDs

(b) Normalized DE for the buffered SNSPD

Figure 4.9: Spectral dependence of the DE for the investigated detectors. The solid lines in (b) are a fit to the spectral dependence to extract the cutoff wavelength. [SIS17]

for large bias currents. No flattening can be seen in the investigated range of λ for the non-buffered SNSPD. In comparison this shows that the λ_C is reached for longer wavelengths on the buffered SNSPD. However, the deterministic regime is not fully reached as it is seen in the bias-dependent measurement (fig. 4.8a). The DE decreases exponentially towards longer wavelengths. The non-buffered SNSPD, on the other hand, shows a non-exponential behavior of the DE towards longer wavelengths. This indicates inhomogeneities inside the detector: constrictions have a higher local current density and remain sensitive for longer wavelengths, whereas most parts of the nanowire are no longer sensitive.

Since the absorption efficiency of the SNSPD for WG illumination is different from the absorption efficiency for top illumination (section 2.3.1), the measured DE cannot be directly transferred to a DE on a WG. To evaluate the general performance, the cutoff wavelength λ_C is extracted at $0.5 \cdot \mathrm{DE}_{\mathrm{plateau}}$. To extract λ_C, the DE(λ)s at different bias levels were normalized on the DE(λ) at $0.95 I_C$. The normalization removes the influence of the setup and the substrate transparency on the efficiency and allows an accurate investigation of the DE at different bias currents in relation to each other. The normalized wavelength dependencies were fitted by an empirical spectral dependence of the DE of SNSPDs [179]:

$$\frac{\mathrm{DE}_2}{\mathrm{DE}_1} = \frac{1 + (\frac{\lambda}{\lambda_{C1}})^{p_1}}{1 + (\frac{\lambda}{\lambda_{C2}})^{p_2}}, \tag{4.1}$$

where DE_1 and λ_{C1} are at the respective DE and λ_C at the bias level to which fig. 4.9b was normalized to. DE_2 and λ_{C2} are the DE and λ_C at the displayed bias level. p describes the power-law dependence of the IDE in the infrared range (details can be found in [179]). The non-exponential dependence of the DE of the non-buffered detector on λ, makes it impossible to get a good agreement of the fit to the data, so only the cutoff for the buffered detector was

Table 4.1: Extracted cutoff wavelengths for a SNSPD with AlN buffer layer at different bias currents.

I_B/I_C	0.95	0.90	0.85	0.8	0.75
λ_C (nm)	796	640	537	455	396

extracted. The fit is displayed in fig. 4.9b as a solid line. The extracted cutoff wavelengths are shown in table 4.1.

At I_B = .95 I_C, λ_C = 796 nm is reached. This is close but nevertheless not enough to be in the deterministic detection regime for the QD wavelength of 900 nm. To reach a cutoff of 900 nm, additional optimizations of the detector are necessary.

4.5 Summary

In this chapter, a double-spiral SNSPD was successfully demonstrated in a design suitable for the integration onto a single-mode WG by utilizing a taper structure. SNSPDs were patternd from NbN with and without AlN. By using the buffer layer the T_C was enhanced from 8.7 K to 9.3 K and the critical current was increased from 8.8 µA to 16.5 µA. However, the thermal coupling was lower, which is in good agreement with previous observations (section 3.3.3 of chapter 3). The higher I_C of the buffered detector led to a higher pulse height and a better separation from electric noise. Several improved parameters for the buffered SNSPD were observed under optical excitation in comparison to the non-buffered SNSDP: The DCR was reduced by three orders of magnitude. The DE was improved over the characterized spectral range from 400 to 1700 nm. At a wavelength of 900 nm, the improvement of the DE is in the order of one magnitude. At an I_B of $0.9 I_C$ and a wavelength of 900 nm (385 nm), a DE of 5 % (31 %) was measured for the buffered SNSPD in comparison to a DE of 0.3 % (5 %) for the non-buffered SNSPD. Since higher pulses reduce the noise contribution to the jitter, the system timing jitter is reduced by 36 ps to 72 ps by the use of the buffer layer. A latching of the buffered detector was observed, whilst no latching was present without buffer layer. The higher latching sensitivity may be caused by the observed weakened thermal coupling of the NbN to the substrate for the buffered samples. For the buffered SNSPD, the cutoff wavelength is shifted towards a longer wavelength and a cutoff wavelength of 796 nm was reached for a $I_B = 0.95 I_C$, which is close to the desired λ_C of 900 nm for the targeted photonic integrated-circuit. However, further optimization is necessary to push the detector into the deterministic regime for a wavelength of 900 nm. In conclusion, the used buffer layer proved effective to increase the performance of a NbN SNSPD on GaAs.

5 SNSPD with Variable Thickness to Minimize the Current-Crowding Effect

The presence of bends with small bending radiik in a superconducting nanowire leads to the presence of a localized high current density region on the inside of the bend [159]. This reduces the critical current of the nanowire and therefore limits the detection efficiency of a SNSPD [143]. In addition, the increased critical current in the bend reduces the entry barrier for thermally activated vortices which are understood as the predominant reason for dark counts [142]. During this thesis, spiral-type detectors were used in chapter 4 and chapter 7 to minimize the current-crowding effect [178] [158]. However, for a PNR detector shown in chapter 6 and for a standard single-loop designs (see section 2.3.8 in chapter 2) for photonic integrated-circuits, a spiral design is not feasible. Furthermore, a meander design has some additional benefits: the resulting detector is sensitive to linear polarization [135], which allows the characterization of the polarization of a light source and enables effective optical coupling of photons to the detector by coupling along its polarization axis. In this chapter, the potential of using a variable thickness design for a SNSPD to reduce the current-crowding effect was studied. SNSPDs were designed and fabricated with nanowires in a standard thickness for SNSPD and thick edges. As a control group SNSPDs were fabricated with uniform thickness and an identical nanowire design. The detectors were studied and compared regarding their superconducting and optical properties. Fabrication and experiments presented in this chapter were done with the support of Michael G. W. Müller [Mül18].

5.1 Current-Density Distribution and Vortex Entry Barrier

So far SNSPD have only been investigated in a designs with uniform thickness. However, the current-crowding effect (see section 2.3.7 in chapter 2) in bends may be removed as a limiting factor for the critical current by adapting the thickness of the SNSPD. The current density in the bends should be reduced by an increase in the bend's thickness in respect to the thickness of the nanowire. If the current density in bends is sufficiently reduced whilst the current-crowding effect is still present, the critical current in the bends remains higher than the critical current of the nanowires and the nanowire can still be efficiently biased. Due to a decrease in the current density, thick bends should have an increased vortex entry barrier, which should decrease the number of dark counts.

A simulated current density and vortex entry barrier distribution for a SNSPD with uniform thickness (U-SNSPD) and for a SNSPD with variable thickness (V-SNSPD) is shown in fig. 5.1. The simulation was done by Saman Jahani, from the California Institute of Technology, for a nanowire width of 100 nm and a bend radius of 25 nm at $I_B = 0.7 I_C$. The thick

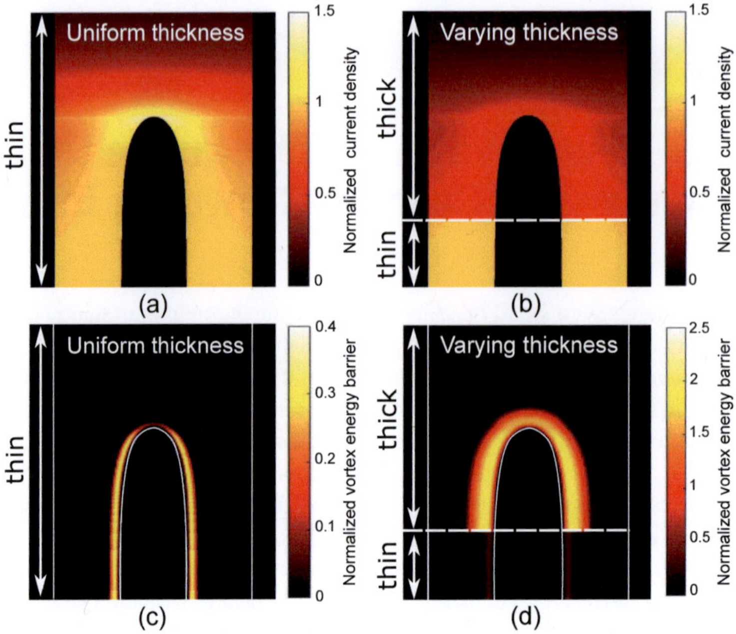

Figure 5.1: Simulated current density (a),(b) and vortex entry barrier (c),(d) for bends with a uniform thickness (a),(c) and with a varying thickness (b),(d). Figure courtesy of Saman Jahani, from the California Institute of Technology.

bends for the V-SNSPD in the simulation are twice as thick as the thin nanowires and the U-SNSPD. In a standard thin bend, the current density inside the bend is increased by a factor of 1.5 (fig. 5.1(a)). This leads to a significant decrease of the vortex entry barrier (fig. 5.1(c)). In comparison, the current density in the thicker bends is significantly decreased (fig. 5.1(b)) and remains smaller than the current density of the nanowire. As a consequence, the vortex entry barrier in the thick bend is significantly increased in comparison to the thin bend and is even higher than the vortex entry barrier in the straight nanowire sections (fig. 5.1(d)). This suggests that the V-SNSPD has a higher I_C and can be biased closer to I_{dep} at a higher saturation level, which increases λ_C and a lowers the amount of dark counts in comparison to the U-SNSPD. To investigate this effect in real devices, corresponding detectors were designed and fabricated.

5.2 Design and Fabrication of Detectors with Variable Thickness

Detectors were designed in a meander design and fabricated using two different fabrication processes, process A and process B. To provide best comparability, U- and V-SNSPDs are made with an identical layout and differ only in the bend thickness and the thickness of the

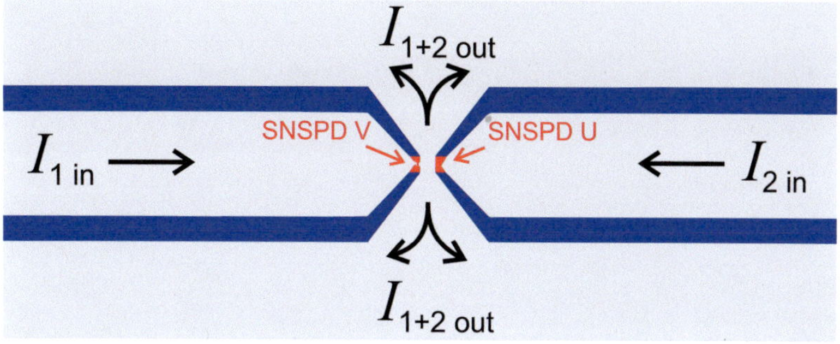

Figure 5.2: Overview drawing of the full chip design. Black arrows visualize the current flow.

nanowires close to the bend. For each sample, a pair of one U- and one V-SNSPDs was implemented, at a distance of 50 µm to each other (fig. 5.2). The close placement allows the exclusion of local influences during fabrication and allows the characterization of both SNSPDs in a single cooling cycle. Both detectors share the same ground plane with an individual coplanar line for readout and biasing for each detector. The resulting current flow is visualized by black arrows. The detectors consist of 19 straight wires with a length of 6 µm and a thickness of ≈5 nm that are connected by bends that complete 180° turns. The thickness of the bends of the V-SNSPD is ≈12 nm and the bends in the U-SNSPD have the same thickness as the nanowire. The meander design was chosen because the current-crowding effect is stronger for sharper turns and, in return, the potential improvement is larger. Detectors fabricated in process A have a design width of 100 nm and those in process B a width of 80 nm. The gap between wires is 60 nm. Next to the SNSPDs, additional dummy nanowires were added to ensure a homogenous width for all wires in the active area after fabrication. Wires without adjacent wires are smaller due to a lower influence of the proximity effect during lithography. In addition, the etching rate of areas differs from the etching rate of gaps which would introduce additional inhomogeneities. To reduce positioning errors between the different patterning steps and to allow a uniform distribution of the current after the thickness change, the thickness of nanowire for the adjacent 500 nm to the bend for V-SNSPD is increased to the bends' thickness. The bend itself has a width of 500 nm, which is significantly larger than the nanowire width to ensure no influence of the outer radius of the bend on I_C. NbN was chosen as detector material since it allows the operation and characterization of SNSPDs at 4.2 K. R-plane sapphire was chosen as a substrate: it has a good lattice matching to NbN, no native oxides and a high mechanical, chemical and thermal stability. This eases the fabrication of high-quality NbN films. The detectors are connected on-chip to RF-coplanar WGs for biasing and readout. In process A, the RF-WG is thinned alongside the nanowire thinning step and has a thickness of ≈5 nm. In process B, the RF-WG is not thinned and has the thickness of the initial NbN film (≈12 nm).

The SNSPD fabrication can be divided into four steps (fig. 5.3): I) film deposition; II) protection of edges and thinning; III) SNSPD patterning and etching; IV) fabrication of RF-coplanar WGs. The SNSPDs that were fabricated using process A were fabricated in the order I)→III)→II)→IV) and the SNSPDs fabricated in process B in the process order

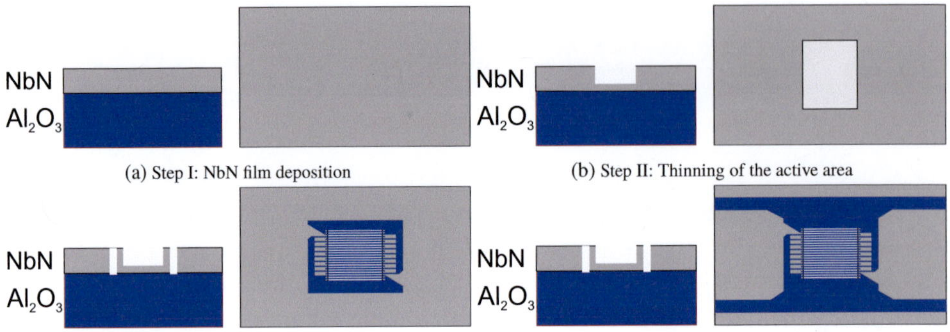

(a) Step I: NbN film deposition

(b) Step II: Thinning of the active area

(c) Step III: SNSPD patterning and etching

(d) Step IV: Coplanar waveguide patterning

Figure 5.3: Illustration of fabrication steps of V-SNSPD.

I)→II)→III)→IV) with slight changes to the process which are explained in the following. For the electron beam lithography process, PMMA 950 K was used at different levels of thickness as positive (exposed pattern is removed) and negative (exposed pattern remains) resist. When electrons hit the PMMA resist, polymer chains in the resist are broken and the exposed resist gets more soluble for a suitable solvent. In a parallel process, the irradiated electrons create a cross-linking of the polymer chains which become hard and insoluble. For a commonly used electron dose during exposure, the polymer breaking predominates and the exposed structure is removed by the use of a developer. However, when the exposure dose is significantly increased, the cross-linking predominates [204]. The typical dose for the negative process is $\approx$100 times higher than the dose needed for the positive process. When the resist is stripped using acetone after exposure with a dose corresponding to the negative process, the unexposed resist is removed and the exposed structure remains. The resist left after the positive process is softer than the resist left after the negative process. The positive resist has an etching rate in the used Ar^+ ion-milling process of $\approx$6.4 nm/min. The resist left after the negative process has a higher contrast [204] and an Ar^+ ion-milling rate of 2 nm/min. Due to the significantly higher required dose, the negative process is very slow and only suitable for the exposure of small areas. In addition, the remaining resist cannot be removed by a standard solvent. For pattern transfer and thinning, Ar^+-milling was used with an etching rate of 1.7 nm/min for NbN.

In the first step, a 12 nm thick NbN film was grown using reactive DC-magnetron sputter deposition (fig. 5.3a).
Process A: the U- and V-SNSPDs were patterned in a positive EBL process using 70 nm of PMMA resist (fig. 5.3c). The pattern was transferred to the NbN film using Ar^+-milling. In a second EBL-process, the bends of the V-SNSPDs were protected by 70 nm of PMMA resist using a negative EBL process. In a second Ar^+-milling step, the full NbN film was thinned to a thickness of 4.7 nm with an exception of the protected edges (V-SNSPD).
Process B: In a first EBL step (fig. 5.3b), windows were patterned into a positive 180 nm thick PMMA950K electron beam resist. For U-SNSPDs, the windows are larger than the detector to thin the full detector including bends. For V-SNSPDs, the windows have a size corresponding to the active area (straight nanowire sections) of the detector. The NbN film in

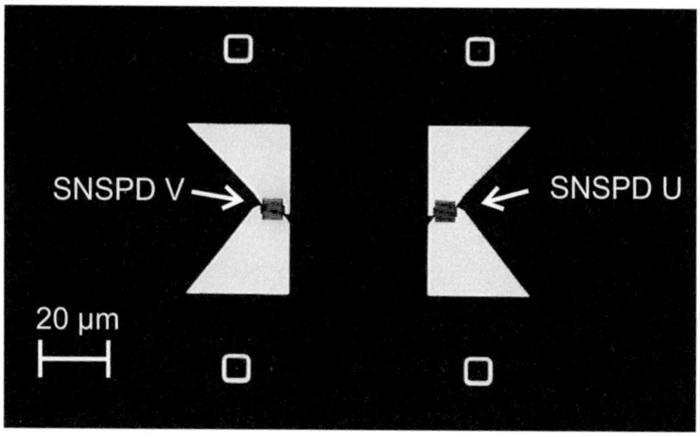

Figure 5.4: Overview SEM image of both detectors after etching, before patterning of the readout. Black areas are NbN, grey areas are sapphire.

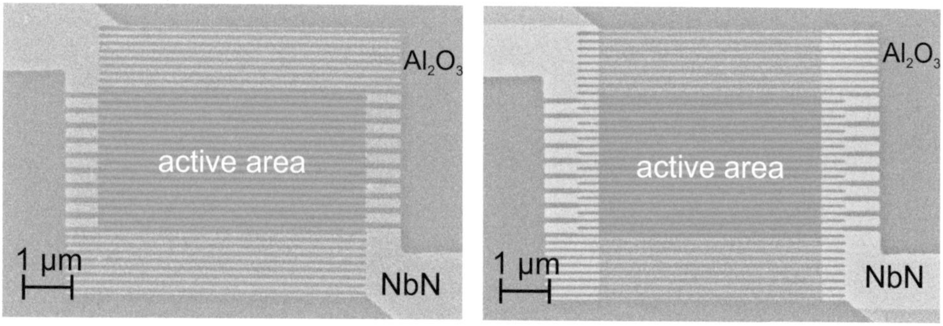

(a) U-SNSPD

(b) V-SNSPD

Figure 5.5: SEM images of the V- (a) and U-SNSPD (b) after fabrication with process A. Light gray shows the thick NbN film, the slightly darker shade is the thinned out SNSPD and the dark gray the sapphire substrate. The active area is highlighted in blue.

the patterned windows was thinned to a thickness of 4.3 nm using Ar^+-milling. The SNSPDs were structured on top of the thinned areas using a two step EBL process: the SNSPDs were patterned in a negative EBL step using 70 nm of resist and the NbN film outside the detector areas was protected by 180 nm of resist structured in a positive EBL process. The SNSPD pattern was transferred to the NbN film by another Ar^+-milling step (fig. 5.3c). A SEM image of a detector pair after fabrication is shown in fig. 5.4.
In a final EBL and Ar^+-milling step for both processes, the coplanar RF-WG for readout and biasing was structured (fig. 5.3d).

The finished detectors after fabrication with process A are shown in fig. 5.5. In the active area, the nanowires have a width of 75 ± 5 nm and a thickness of 4.7 nm with a gap of 85 ± 5 nm. This corresponds to a filling factor of 47 %. Close to the bend, the width of the V-SNSPD is increased to a width of 110 ± 5 nm. The bends of the V-SNSPDs have a thickness of 12.1 nm. The size of the active areas of the U-SNSPDs is $3\,\mu m \times 6\,\mu m$, and $3\,\mu m \times 5\,\mu m$ for the V-SNSPDs.

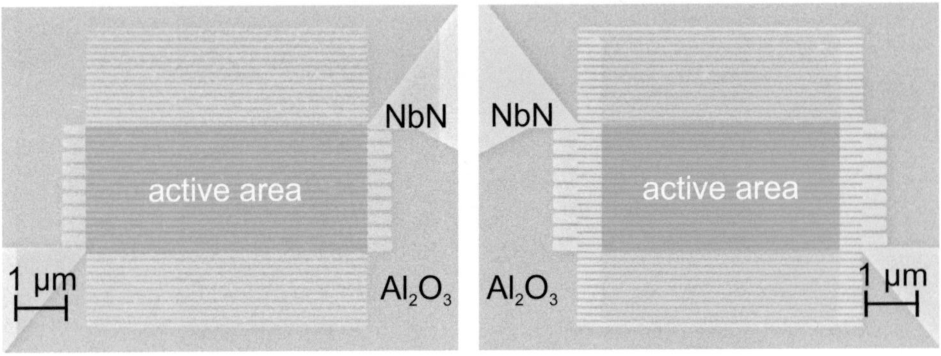

(a) U-SNSPD

(b) V-SNSPD

Figure 5.6: SEM images of the V- (a) and U-SNSPD (b) after fabrication with process B. Light gray shows the thick NbN film, the slightly darker shade is the thinned out SNSPD and the dark gray the sapphire substrate. The active area is highlighted in blue.

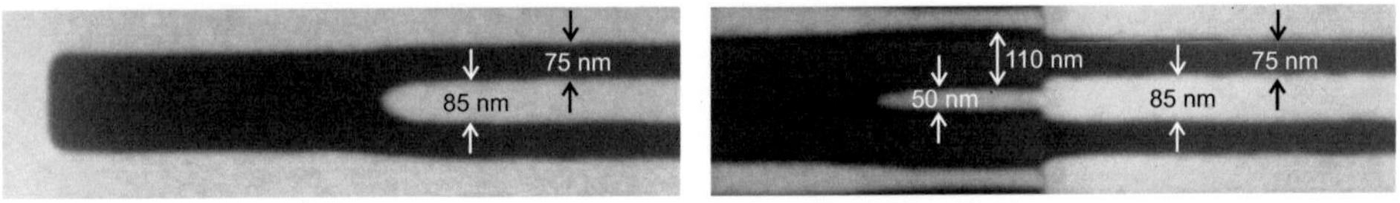

(a) Single U-SNSPD bend after fabrication with process A

(b) Single V-SNSPD bend after fabrication with process A

Figure 5.7: SEM of a single bend after fabrication with process A. The thinning process decreases the width, which causes a significantly smaller bend radius in the V-SNSPD in comparison to the U-SNSPD.

SEM images of detectors after fabrication with process B are shown in fig. 5.6. The nanowires have a width of 85 ± 5 nm and a thickness in the active area of 4.3 nm at a filling factor of 60 %. The bends of the V-SNSPDs have a thickness of 11.7 nm. The size of the active areas of the U-SNSPDs is 2.6 µm × 6 µm, and 2.6 µm × 5 µm for the V-SNSPDs.

The bends after fabrication with process A (fig. 5.7) compared to bends fabricated with process B (fig. 5.8) reveal significant geometrical differences due to the different order of fabrication processes and the different use of positive and negative resist. After process A, the bend is rounded, while the bends after process B consist of two almost perfect 90° turns. The thinning of the active area after the SNSPD patterning in process A leads to a width reduction of the nanowire by 35 nm. For the V-SNSPD, the protection of bends prevent a nanowire thinning close to the bend and, consequently, the width reduction effect of the etching. Hence, an abrupt change in width is visible for the transition from the thinned to the not thinned region (fig. 5.7b). As a consequence, the bend radius of the U-SNSPD (fig. 5.7b) is ≈40 nm and hence, significantly larger than the bend radius of the V-SNSPD (≈25 nm). It follows that the smaller radius of the bend increases the current crowding while the larger thickness may reduce the current crowding. This makes an evaluation of the influence of a variable thickness on the current-crowding effect difficult. The SNSPDs fabricated with process B are not influenced by this effect since the thinning step is performed prior to the SNSPD fabrication. The width of the nanowire remains constant, even after the thickness change. The

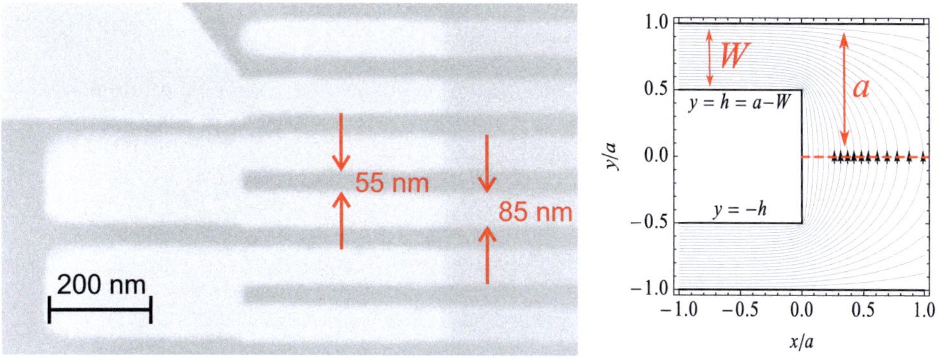

(a) V-SNSPD bends after fabrication with process B

(b) Theoretical current distribution

Figure 5.8: (a) SEM of detector bends after fabrication with process B. (b) Simulated current distribution for a bend consisting of two perfect 90° turns. (b) was reprinted with permission from [143]. Copyright 2011 by the American Physical Society.

almost perfect 90° turns for the process B can be explained by the different usage of a negative and positive process during fabrication. While the SNSPD in process A was patterned using positive resist and the bends were protected using negative resist, the opposite was done for process B: the SNSPD was protected by the negative resist and the bends were protected by the positive resist. After exposure and development due to the proximity effect in electron beam lithography [205], the edge profile of the resist is not perfectly vertical but has an inward (positive-process) or outward (negative-process) slope. The Ar^+ milling in combination with a slope cause the resist to slowly be removed from the side during etching. This slow removal process introduces a slope in the etch profile of the NbN pattern. This slope is smoother for the positive resist due to the larger etching rate. As a result, the negative process has a higher resolution. In addition, the transition from the thinned nanowires to the not thinned bends is smoother when a positive resist is used to protect the edges. In case of the negative resist used for edge protection in process A, a steeper step exists from the thinned nanowire to the not thinned edges, which might introduce further current crowding and may lead to an additional suppression of the critical current in the V-SNSPDs.
The almost perfect 90° turns and the homogeneity in width after process B allow an estimation of the current-crowding effect using [143]. The reduction of the critical current, R, by the current-crowding effect in a bend with two perfect 90° (fig. 5.8b) can be calculated as:

$$R = \frac{3}{2}\left(\frac{\pi\xi_{GL}\sqrt{1-(W/a)^2}}{2W}\right)^{\frac{1}{3}}, \tag{5.1}$$

where W is the width of the nanowire and $a = W + \frac{g}{2}$ where g is the width of the gap (see: fig. 5.8b). Using the measured $W = 86\,\text{nm}$ and $a = 113\,\text{nm}$ and a typical value of the Ginzburg-Landau coherence length ξ_{GL} for NbN of 5 nm [131], a $R = 0.58$ is calculated. Out

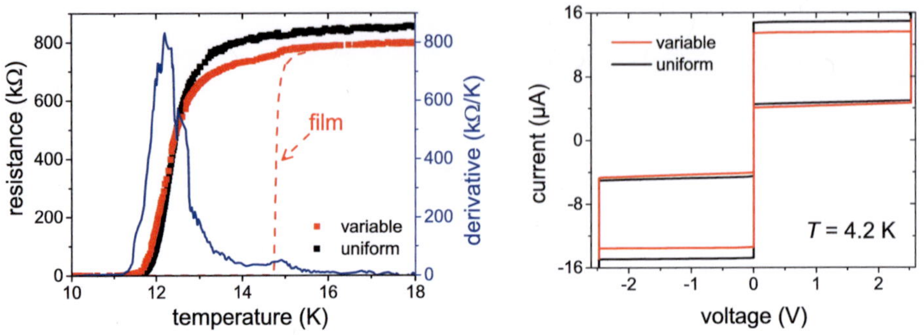

(a) $R(T)$ characteristic

(b) $I(V)$ characteristic

Figure 5.9: Comparison of superconducting characteristics of the V-SNSPD (red) and the U-SNSPD (black) fabricated in process A. The data shown was measured using detector pair D1-A (table 5.1). (a) Resistance-temperature dependence for both detectors (squares). For comparison, the red dashed line shows the $R(T)$ characteristic of the film with its maximal resistance normalized to the maximal resistance of the V-SNSPD. The blue solid line shows the derivative of the RT-dependence of the V-SNSPD. (b) Current-voltage characteristic of both detectors at a temperature of 4.2 K. The vertical dependence at ±2.5 V is caused by a saturation of the used ADC in the measurement system.

of this R, an improvement of I_C of $R^{-1} - 1 = 71$ % is expected after fabrication with process B, if the current crowding is fully removed by the variable thickness approach.

5.3 Superconducting Properties

The superconducting properties of detector pairs were investigated by measuring the $R(T)$-dependence and $I(V)$-dependence. In this section, results from detector pairs fabricated using process A are discussed, followed by a discussion of detector-pairs fabricated using process B and a comparison of the results for both detector types.

SNSPDs Fabricated with Process A

The $R(T)$ characteristic of a detector pair fabricated using process A (D1-A) is shown in fig. 5.9. The U-SNSPD shows a single superconducting transition, while a two step transition is visible for the V-SNSPD (the derivative of the transition of the V-SNSPD is shown in blue). Each displayed data point is a average of 10-points, to reduce excessive noise. A first step at ≈14.5 K with a corresponding increase of the derivative coincides with the superconducting transition of the 12.1 nm thick, initially grown film and can be contributed to the superconducting transition of the thick bends. The second step coincides with the transition of the U-SNSPD and can be contributed to the thin nanowire sections and RF-WG (4.7 nm). The $I(V)$ characteristic of both SNSPDs is similar, with a higher I_C for the uniform detector. In table 5.1 an overview of extracted superconducting parameters for all operating detector pairs fabricated with process A is given.

Table 5.1: Superconducting parameters for the SNSPD pairs fabricated in process A. Detectors from one pair are indicated by the same number, followed up by an U for the uniform and a V for the variable thickness SNSPD.

Sample	T_C	ΔT_C	R_N	$I_C(4.2\,K)$	$I_r(4.2\,K)$	$I_C^{GL}(0\,K)$	$\frac{I_C^{GL}(0)(V)}{I_C^{GL}(0)(U)}$
	K	K	kΩ	µA	µA	µA	
Film-A	14.7	0.3					
D1V-A	11.2	1.9	810	13.4	4.1	40.1	0.92
D1U-A	11.4	1.4	860	14.8	4.6	43.7	
D2V-A	11.3	1.4	760	12.8	4.1	38.0	0.82
D2U-A	11.3	1.6	820	15.6	4.7	46.4	
D3V-A	10.8	1.8	940	9.4	3.1	28.5	0.88
D3U-A	11.1	1.5	1000	10.8	3.8	32.5	

The T_C and ΔT_C of detectors is similar for both types. The thinning step reduces the T_C by $\approx$3.5 K. Due to the lower normal resistance of the thick edges, the normal resistance of the V-SNSPD is $55 \pm 5\,k\Omega$ lower than the normal resistance of the U-SNSPD. The measured I_C and I_r are lower for the V-SNSPD. Since the detectors have a slightly different T_C, the respective I_C were normalized for better comparison by estimating the I_C for $T \mapsto 0\,K$ to decrease the influence of a variation in T_C on the measured I_C. The normalization was done by calculating $I_C^{GL}(0)$ using the Ginzburg-Landau temperature dependence of the critical current in combination with the Kupriyanov-Lukichev correction factor $KL(T)$ [206] [207] eq. (5.2):

$$j_C(T) = j_C^{GL}(0) \cdot \left(1 - \frac{T}{T_C}\right)^{\frac{3}{2}} \cdot KL(T), \tag{5.2}$$

where j_C is the critical current density. $KL(T)$ for a superconductor in the dirty limit (the mean free path length for electrons $l_e << \xi_{GL}$) can be extracted out of the analytical approximation to figure 1 in [206] eq. (5.3):

$$KL(T) \approx 1.87 - \frac{1.46}{1 + (0.7 \cdot \frac{T}{T_C})^{1.12}}. \tag{5.3}$$

$I_C^{GL}(0)$ is then calculated using eq. (5.2) in combination with eq. (5.3) and the proportionality $I_C \propto j_C$:

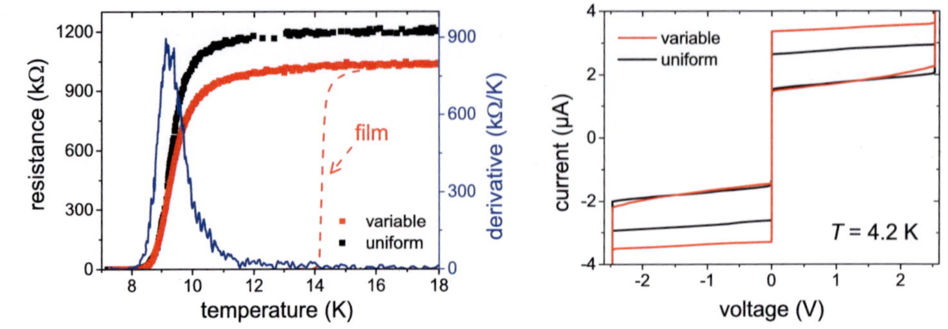

(a) $R(T)$ characteristic

(b) $I(V)$ characteristic

Figure 5.10: Comparison of the superconducting characteristics of the V-SNSPD (red) and the U-SNSPD (black) fabricated with process B. The shown data was measured using detector pair D1-B (table 5.2). (a) Resistance-temperature dependence for both detectors (squares). For comparison, the red dashed line shows the $R(T)$ characteristic of the film with its maximal resistance normalized to the maximal resistance of the V-SNSPD. The blue solid line shows the derivative of the RT-dependence of the V-SNSPD. (b) Current-voltage characteristic of both detectors at a temperature of 4.2 K. The vertical dependence at ±2.5 V is caused by a saturation of the used ADC in the measurement system.

$$I_{\mathrm{C}}^{\mathrm{GL}}(0) = I_{\mathrm{C}}(4.2\,\mathrm{K}) \cdot \left(1 - \frac{4.2\,\mathrm{K}}{T_{\mathrm{C}}}\right)^{-\frac{3}{2}} \cdot \left(1.87 - \frac{1.46}{1 + (0.7 \cdot \frac{4.2\,\mathrm{K}}{T_{\mathrm{C}}})^{1.12}}\right)^{-1}. \tag{5.4}$$

The improvement of I_{C} was calculated out of $\frac{I_{\mathrm{C}}^{GL}(0)(V)}{I_{\mathrm{C}}^{GL}(0)(U)}$. On average, $I_{\mathrm{C}}^{GL}(0)$ is reduced by $13 \pm 5\,\%$ for the V-SNSPD. The reduction of the critical current by the smaller bend radius of the V-SNSPD (25 nm) in comparison to the bend of the U-SNSPD (40 nm) overweight the effect of the thicker bends.

SNSPDs Fabricated with Process B

The resistance-temperature dependence for a detector pair (D1-B) fabricated using process B is shown in fig. 5.10a. The blue solid line shows the derivative of the transition of the V-SNSPD after averaging over 10-points to smooth the curve. For both detector types, the progression of the $R(T)$ characteristics are comparable. The normal state resistance R_N of the uniform detector is higher since the thinned bends have a higher resistance. In contrast to SNSPDs fabricated using process A, only one transition step is clearly visible for the V-SNSPD and no additional step can be extracted out of the derivative. The $I(V)$-dependence (fig. 5.10b) shows a similar progression for both detectors and a higher I_{C} of the V-SNSPD.

An overview of extracted parameters for detector pairs and the unstructured NbN film can be found in table 5.2. The T_{C} is slightly higher for the V-SNSPD. The processing and thinning

Table 5.2: Superconducting parameters of SNSPD pairs fabricated in process B.

Sample	T_C	ΔT_C	R_N	$I_C(4.2\,\mathrm{K})$	$I_r(4.2\,\mathrm{K})$	$I_C^{GL}(0\,\mathrm{K})$	$\frac{I_C^{GL}(0)(V)}{I_C^{GL}(0)(U)}$
	K	K	kΩ	µA	µA	µA	
Film	13.9	0.4					
D1V-B	8.0	2.0	1040	3.4	1.5	7.8	1.28
D1U-B	8.0	1.5	1210	2.6	1.6	6.1	
D2V-B	8.6	3.0	680	5.1	2.0	10.4	1.32
D2U-B	8.0	1.5	990	3.4	2.2	7.9	
D3V-B	8.9	2.2	810	4.9	2.4	9.4	1.32
D3U-B	7.8	1.7	1110	2.9	2.0	7.1	

of the active detector area (11.7 nm → 4.3 nm) reduces the T_C by ≈5.5 K. The ΔT_C for the V-SNSPD is larger, which can be caused by the different T_C of the thick bends in comparison to the nanowire. Whilst all V-SNSPDs have a higher I_C than their uniform counterparts, the retrapping current I_r is similar for both detector types. The R_N of the U-SNSPD is significantly higher than of the V-SNSPD due to the lack of thick edges. For better comparison of I_C of the detectors despite the differences in T_C, the previously introduced $I_C^{GL}(0)$ was calculated. The calculated $\frac{I_C^{GL}(0)(V)}{I_C^{GL}(0)(U)}$ reveals an average $I_C^{GL}(0)$ improvement of 30 ± 2 % for the V-detector. The difference of the experimental improvement of 30 ± 2 % compared to the theoretical improvement of 70 % can have several reasons: albeit weaker, current crowding may still be present in the thicker bends or the critical current may already be suppressed by imperfections in the nanowire for the V-SNSPD. In addition, the effect of current crowding in the U-SNSPD may be lower than expected, since the fabricated bends do not have perfectly sharp 90° turns. Using equation (63) and figure 3 out of [143] an effective bend radius of ≈5 nm ($\approx 1 \cdot \xi_{GL}$ with $\xi_{GL} = 5\,\mathrm{nm}$ [131]) is enough to fully explain the observed improvement. This radius is plausible for the remaining radius as seen in the SEM image of the bend in fig. 5.8a within the resolution of the image.

Comparison

In comparison, the difference in T_C (≈3 K) and I_C (≈10 µA) between SNSPDs fabricated in process A and process B is stronger than expected for a difference in thickness of 0.5 nm (10 %) of the tinned film. The T_C reduction of the process B by the thinning process is significantly stronger than for process A (≈5.5 K versus ≈3.5 K). Also, the experimental I_C after process B is over three times lower than the I_C values after process A. In comparison to typical I_C values of a NbN meander SNSPD on sapphire of ≈30 µA [208], the I_C of SNSPDis lower by a factor of two after process A and by a factor of ten lower after process B. This might be explained by the influence of the etching process on the thin NbN film. The Ar-ions from the Ar^+-milling process not only etch the material physically from the top. If they do

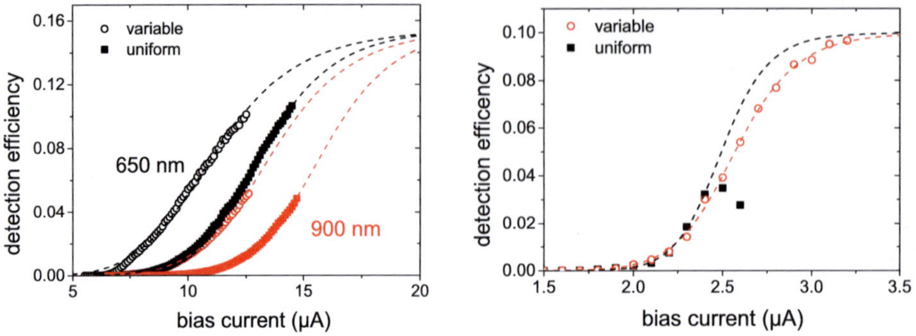

(a) SNSPDs fabricated with process A

(b) SNSPDs fabricated with process B

Figure 5.11: Bias dependence of the DE of V- (open circles) in comparison to U-SNSPDs (closed squares). The dashed lines are sigmodial logistic fits to the data. (a) Detector pair fabricated in process A. The bias dependence is shown for $\lambda = 650\,\mathrm{nm}$ in black and for $\lambda = 900\,\mathrm{nm}$ in red. (b) Detector pair fabricated in process B. The bias dependence is shown for $\lambda = 500\,\mathrm{nm}$.

not collide with an atom on the surface, they penetrate deep into the film and can disturb the crystalline structure of the NbN. A direct comparison of the T_C for a film grown to a thickness of 5.5 nm and a film milled to 5.5 nm from 12 nm showed a suppression of T_C by more than 2 K connected to the etching process [Mül18].
While SNSPDs fabricated using process A showed a decrease of $I_C^{GL}(0)$ by $13 \pm 5\,\%$, an increase of $30 \pm 2\,\%$ was observed for SNSPDs fabricated using process B. This proves the importance of controlling the bend radius for a SNSPD with variable thickness. In addition, a steeper transition from the thinned nanowires to the thicker bends caused by the fabrication process A may have introduced additional current crowding for the V-SNSPD.

5.4 Optical Properties

The photo-response of the detectors was characterized in a fiber-coupled dipstick setup (section 4.3 of chapter 4) using top excitation at a bath temperature of 4.2 K. One detector pair was characterized for each fabrication process regarding their DE and DCR. The detection efficiency was calculated as $\mathrm{DE} = \frac{\mathrm{CR}-\mathrm{DCR}}{N_{Ph}}$, where N_{Ph} is the number of photons on the active area of the respective detector. For process A, the detector pair D1-A was chosen for characterization. For process B, the detector-pair D1-B was chosen for characterization to exclude an influence of the T_C in the measured result.

The DE was investigated in relation to I_B for the SNSPDs fabricated in process A (fig. 5.11a) at $\lambda = 650\,\mathrm{nm}$ and at $\lambda = 900\,\mathrm{nm}$ and at $\lambda = 500\,\mathrm{nm}$ for the SNSPDs fabricated in process B (fig. 5.11b). To extract the level of saturation for the internal efficiency, the data was fitted with a logistic sigmoid function. For the detector pair fabricated using process A, the dependence on the bias current is similar for the U- and the V-SNSPD. However, the dependence is shifted by $2\,\mu\mathrm{A}$ towards a higher bias current for the U-SNSPD. Both detectors reach a maximal

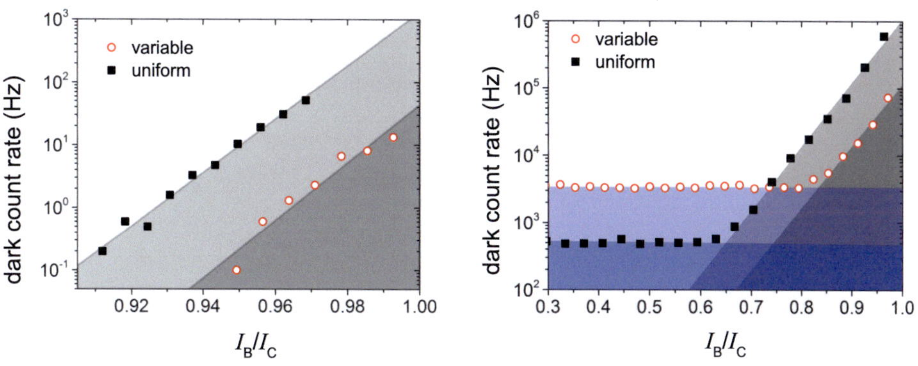

(a) SNSPDs fabricated with process A

(b) SNSPDs fabricated with process B

Figure 5.12: Bias dependence of the dark count rate. The U-SNSPD is shown as a black square and the V-SNSPD is shown as a red circle. (a) DCR dependence of the SNDPDs with fabrication process A. (b) DCR dependence of the detector pair fabricated using process B. The colorized areas visualize the contribution of electrical noise (blue) and intrinsic dark counts (gray).

detection efficiency of ≈10 % at 650 nm and of ≈5 % at 900 nm. The extracted saturation is ≈67 % (≈69 %) at 650 nm for the V-SNSPD (U-SNSPD) and ≈34 % (≈31 %) at 900 nm. Both detectors fabricated in process B start to detect photons at a bias current of ≈2 µA. The dependence is comparable for both detectors until the I_C of the U-SNSPD is reached. The DE of the V-SNSPD further increases until it starts to level out at a DE close to 10 %. The almost identical dependence for both detectors indicates a high similarity of straight nanowires of both SNSPDs with a suppressed I_C for the U-SNSPD. U-SNSPD reaches 32 % of saturation whilst the V-SNSPD almost reaches full saturation with a saturation level of 97 %.

The DCR for both SNSPDs with both detector types is shown in fig. 5.12 in dependence of I_B/I_C. For SNSPDs fabricated in process A, an exponential dependence is measured. The DCR of the U-SNSPD is hereby one order of magnitude higher than the DCR of the V-SNSPD. For SNSPDs fabricated in process B, the DCR is constant at low bias currents and higher for the V-SNSPD. The DCR starts to rise exponentially at $I_B \approx 0.65 I_C$ for the U-detector and at $I_B \approx 0.65 I_C$ for the V-detector. The dark counts were fitted with an offset exponential function to separate electronic noise from intrinsic dark counts. The offset is caused by electronic noise and is connected to a low SNR of detector pulses due to the low I_C of the investigated detectors. The intrinsic DCR of detectors is seen in the exponential dependence close to the respective I_C (see section 2.3.3 in chapter 2). While the electric noise during measurement of the V-SNSPD was higher, the intrinsic DCR is one order of magnitude lower for the same relative bias current. In direct comparison, the DCR of the detectors fabricated in process B is significantly higher than of detectors fabricated in process A. The higher DCR may be connected to a larger number of defects in detectors fabricated with in process B caused by the thinning process.

The spectral dependence of the DE was measured at $I_B = 0.85 I_C$ (blue) and $0.9 I_C$ (red) for λ from 500 to 1700 nm (fig. 5.13). For the SNSPD-pair fabricated in process A, the DE for a short λ is comparable for both detectors. However, towards longer λ the decline of

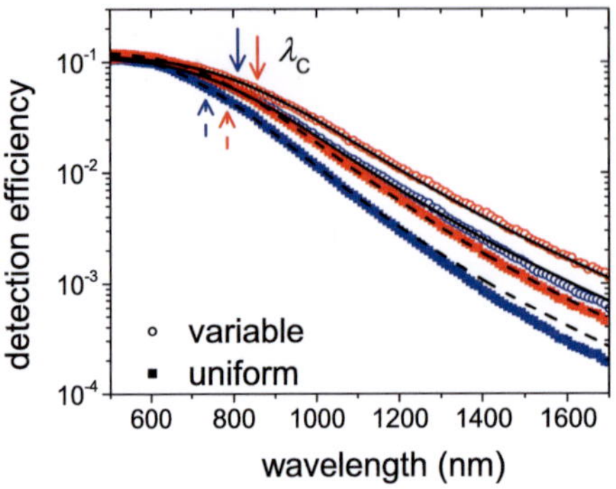

(a) SNSPDs fabricated in process A

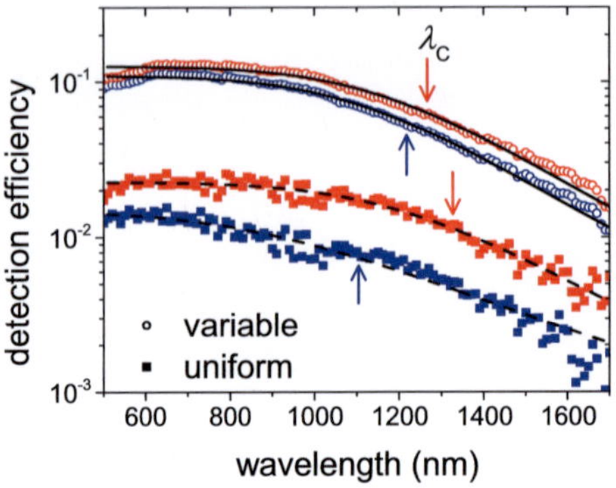

(b) SNSPDs fabricated in process B

Figure 5.13: Spectral dependence of the DE at $I_B = 0.85 I_C$ (blue) and $0.9 I_C$ (red) for the V-SNSPDs in dots and U-SNSPDs in squares. The lines are fits of the data with eq. (5.5) to extract λ_C for the V-SNSPDs (solid) and the U-SNSPDs (dashed). The arrows indicate the positions of the extracted λ_C.

Table 5.3: Extracted λ_C values out of a fit of eq. (5.5) to the spectral dependence fig. 5.13.

	U-SNSPD		V-SNSPD	
Device	$\lambda_C(0.85 I_C)$	$\lambda_C(0.90 I_C)$	$\lambda_C(0.85 I_C)$	$\lambda_C(0.90 I_C)$
D1-A	734 nm	785 nm	810 nm	858 nm
D1-B	1104 nm	1326 nm	1218 nm	1262 nm

the DE for the V-SNSPD is less pronounced than for the U-SNSPD. For the SNSPD-pair fabricated in process B in the investigated spectral range, the DE is significantly higher for the V-SNSPDs. For the V- as well as the U-SNSPD, a plateau of the DE ($DE_{plateau}$) can be seen for shorter wavelengths followed by a roll off. The absolute DE value and the slight increase of DE from $\lambda = 500$ to 620 nm for the V-detector is in good agreement with the calculated absorption efficiency in [135] for orthogonal polarized light in respect to the meander lines. λ_C at $0.5 \cdot DE_{plateau}$ can be extracted by an empirical fit to the spectral dependence of the DE with [157]:

$$DE(\lambda) = DE_{plateau} \cdot \left(1 + \left(\frac{\lambda}{\lambda_C}\right)^p\right)^{-1}, \tag{5.5}$$

where p describes the power-law decrease of efficiency towards the infrared. $DE_{plateau}$, λ_C and p were used as free fit parameters. The extracted values for λ_C are collected in table 5.3. For all detectors, except for the U-SNSPD fabricated in process B, a $DE_{plateau}$ of 12 ± 2 % is reached.

For the detector pair fabricated in process A, a larger λ_C is reached with the V-SNSPD. With an increase of the relative bias current from 0.85 to $0.90 I_C$, λ_C is shifted towards longer wavelengths by 50 ± 2 nm for both detectors. Since the bias dependence for both detectors shows a very similar DE-current relation, despite being shifted by ≈2 μA, the observed differences of λ_C, I_C and I_r of SNSPDs fabricated in process A might be related to a slightly different width of the nanowire for the U-SNSPD in comparison to the V-SNSPD. Using the linear relation of λ_C to the wire width w of the diffusion-based hotspot model (section 2.3.2 of chapter 2) with the assumption that both detectors are biased to the same relative I_B/I_{dep} and have a comparable thickness, identical temperature and the same material parameters, the relation of λ_{C_1} to λ_{C_2} of the SNSPDs can be expressed in relation to the nanowire widths w_1, w_2 as:

$$\lambda_{C_1} = \frac{w_2}{w_1} \lambda_{C_2}. \tag{5.6}$$

Using eq. (5.6), the difference in λ_C at the investigated bias levels can be explained by a 7 nm smaller width of the V-SNSPD (68 nm versus 75 nm). A width difference of 7 nm also explains the differences in I_C and I_r. Current density (j) values, calculated using a width difference of 7 nm, show a convincing coincidence:

$$j_{C,\text{V-SNSPD,corrected}} = j_{C,\text{U-SNSPD}} = 4.2\,\text{MA/cm}^2 \tag{5.7}$$

$$j_{r,\text{V-SNSPD,corrected}} = j_{r,\text{U-SNSPD}} = 1.3\,\text{MA/cm}^2. \tag{5.8}$$

Hence, the differences of I_C, I_r, DE and λ_C can be explained by a narrower width of the nanowire of the V-SNSPD in comparison to the U-SNSPD. However, this width differential does not explain the larger observed DCR of the U-SNSPD. Whilst the smaller bend radius seen for the V-SNSPD or the abrupt thickness transition seem to nullify an effect of the thicker bends on the I_B/I_{dep}, a constant j_C for a smaller width is already a beneficial effect of the thicker bends on current crowding. Usually, a decrease in j_C is observed for a smaller nanowire width [207]. In addition, the amount of observed dark counts is reduced for the same I_B/I_C.

For SNSPDs fabricated in process B, despite a one order of magnitude lower DE of the U-SNSPD, the extracted λ_C (table 5.3) is larger at $I_B = 0.90 I_C$ in comparison to the V-SNSPD. However, the relative decrease in DE and the decrease in λ_C is significantly stronger for the U-SNSPD when the I_B is decreased from 0.90 to $0.85 I_C$. Overall, the extracted values of λ_C are large for meander NbN SNSPD at a temperature of 4.2 K and comparable with the λ_C observed for a single-spiral SNSPD [158]. The long plateau of the U-SNSPD is unexpected. It is generally understood that for wavelengths where the DE is in a plateau SNSPDs operate in the deterministic detection regime with an IDE close to unity and a further increase of photon energy does not yield a higher DE of the detector since it operates in its absorption limit [209]. While the V-SNSPD, seem to operate in the deterministic detection regime in the plateau, the low DE and the not saturated bias dependence (fig. 5.11b) for the U-SNSPD convincingly show that the main share of the nanowire is not in the deterministic detection regime at the applied bias current. The presence of the plateau might be explained by a high number of defects due to the thinning in the film. With a locally suppressed I_{dep} due to defects, single parts of the nanowire might already be in the deterministic region, whilst the main share of the detector is under-biased and does not contribute to the detection of photons. Since both detectors differ in their bend thickness, the observed effect may also be explained by detection of photons in bends of the U-SNSPD: at the used bias current, the U-SNSPD may only be efficiently biased in the detector edges due to current crowding. In contrast, the straight parts of the nanowire are not sufficiently biased and the active area is significantly smaller which explains the low DE of the U-SNSPD in comparison to the DE of the V-SNSPD. In conclusion, the significant improvement of the DE in the order of one magnitude indicates a more efficient biasing of the V-SNSPD with a dark count rate in the order of one magnitude lower. The high coincidences of the superconducting parameters and the coinciding bias dependence (fig. 5.11b) indicate a high comparability of both nanowires. From the comparable design of the turn, apart from the thickness, it can be concluded that the observed improvements stem from a less strong current suppression and a larger vortex entry barrier in bends, which confirms the expected behavior.

When comparing SNSPDs fabricated in process A with SNSPDs fabricated in process B, the significant larger λ_C for SNSPDs fabricated in process B, despite a significantly lower critical current, meets the eye. A stronger destruction of the crystalline structure of the NbN film, caused by the longer etching process, may significantly suppress the superconductivity in the film. Homogeneously spatially distributed defects may facilitate a detector in which I_C is suppressed in the full detector and which can hence still be sufficiently biased with a significantly lower absolute bias current. With a sufficiently suppressed energy gap in the wire, it will get more sensitive to lower energy photons. In addition, a present high defect density may significantly lower the vortex entry barrier which could explain the large absolute number of observed dark counts in SNSPDs fabricated using process B (fig. 5.12b). This suggests the use of a deterministic destruction of the crystalline structure of the NbN film for SNSPDs with a large λ_C at the cost of an increased DCR. The $DE_{plateau}$ of $12 \pm 2\,\%$ of detectors is roughly half the *DE* that can be observed for NbN SNSPD close to saturation [208] but is in good agreement with the absorption for a single polarization for a NbN film with 4.6 nm thickness [135].

5.5 Summary

A new approach to avoid current crowding using a variable thickness was proposed. SNSPDs were fabricated as meanders with 180° turns in the bends by using two different fabrication processes. Detectors where fabricated in pairs, with one detector having thin nanowires and thick bends (variable) and a detector with uniform thickness (uniform) as a control. The variable SNSPDs have a thickness of $\approx$5 nm in the wires and of $\approx$12 nm in the bends and the uniform SNSPD control devices have a thickness of $\approx$5 nm in the wires and bends. The variable thickness was achieved by deterministic thinning of an initial $\approx$12 nm thick NbN film by Ar^+-ion milling. In detectors fabricated in process A, the SNSPDs were patterned prior to the thinning process; detectors fabricated in process B were first thinned and the SNSPDs were patterned afterward. The thinning process after SNSPD patterning led to a decrease in nanowire width in the thinned areas. This caused a significantly smaller bend radius of the variable SNSPD (25 nm), in comparison to the bend radius of the uniform SNSPD (40 nm) for SNSPDs fabricated in process A. This width reduction was not present for the changed order during the fabrication process B. The etching process for the thinning altered the film quality by significantly decreasing the critical temperature and current of the film. Due to the different bend radii and a 7 nm smaller width, for SNSPDs fabricated in process A the critical current was reduced by 13 ± 5 % for the detector with variable thickness. The smaller bend radius in the variable thickness detector, counteracts the effect of the larger thickness. Nevertheless, due to an identical critical current density at a smaller width, the λ_C was increased by ≈70 nm for the same relative bias current. For the SNSPDs fabricated in process B, which is characterized by a homogeneous wire width and identical bend radii for both detector types, the critical current was increased by 30 ± 2 % using a variable thickness. The difference to a theoretically predicted increase of 71 % for the critical current can be explained by not perfectly sharp turns with a bend radius of ≈5 nm. Connected to the increased critical current, the variable detector could be operated at 97 % saturation at $\lambda = 500$ nm whilst the uniform SNSPD was limited to 32 %. This allowed for a significant increase in DE over the full investigated spectral range and a $\lambda_C = 1260$ nm at $I_B = 0.90 I_C$, which is comparable with the λ_C obtainable for NbN single-spiral SNSPDs [158]. For both fabrication processes, the dark count rate was reduced by one order of magnitude for the V-SNSPD, which proves an increased vortex entry barrier in the bends. In conclusion, a variable-thickness design for SNSPDs is effective to reduce the influence of current crowding for a given bend geometry. This can be very useful for the creation of SNSPD arrays, high efficient detectors with polarization sensitivity and waveguide SNSPDs. In addition, it was found that a deterministic degradation of the film may be suitable to increase the λ_C of a SNSPD.

6 Ultra-fast Photon-Number Resolving Detectors

SNSPDs generally allow the detection of only one photon at a time, however, photonic quantum computers using boson sampling [210] or the KLM-scheme [13] require the direct characterization of photon-number states (Fock states). This can either be done by using one beam splitter and one single-photon detector for each individual number state or by using a single photon-number resolving detector with a sufficient photon-number resolution (PNR). For linear optical quantum computing (LOQC) a PNR of 4 is required. In addition, quantum receivers for coherent communication greatly benefit from a 3-photon resolution. With a 3-photon resolving detector, the error rate during communication can be improved by 27 dB in comparison to the quantum limit of a classical receiver [89]. Whilst series photon number resolving SNSPD (PNR-SNSPD) have already been demonstrated and have been studied extensively [102] [105] [104] [103] [40] [169], they have so far been demonstrated only in fiber-coupled setups at temperatures below 3 K and only on GaAs-substrate. In this chapter, series PNR-SNSPD made from NbN films with palladium (Pd) resistors were investigated for two distinct applications. PNR-SNSPD were investigated on a sapphire substrate for telecommunications and on GaAs for waveguide integration into quantum-photonic integrated circuits. Both detectors were investigated using a standard SNSPD setup at temperatures ≥4 K regarding their PNR properties and their ability to represent the statistics of a light source with high accuracy. A scalable detector design suitable for top illumination and waveguide integration is realized. The first series PNR-SNSPDs with 3-photon resolution on sapphire were fabricated and characterized regarding their optical properties focused around a telecommunication wavelength of 1550 nm. Series PNR-SNSPDs on GaAs with a waveguide optimized design were fabricated with 4-photon resolution and investigated over a broad spectral range from 600 to 1700 nm on the single-photon level and for four discrete wavelengths from 740 to 1550 nm regarding their PNR characteristic. The accuracy of the measured PNR-characteristic was studied for ultra-fast pulsed lasers with a pulse repetition frequency (PRF) of up to 100 MHz. To demonstrate the utility of a PNR detector for quantum-photonic integrated-circuits (QPICs), the ability of such a detector to discriminate a light source with a Poissonian statistics from a light source with a sub-Poissonian statistics is demonstrated in a free space optical setup.

Parts of the work presented in this chapter are published in IEEE Transactions on Applied Superconductivity under the title: *"Characterization of a photon-number resolving SNSPD using Poissonian and sub-Poissonian light"* [SRS$^+$19]. The work presented in this chapter was done with support of E.Reutter [Reu17]. All optical experiments performed in the later described Setup 1 were supported by M. Schwartz and J. Binz from the University of Stuttgart.

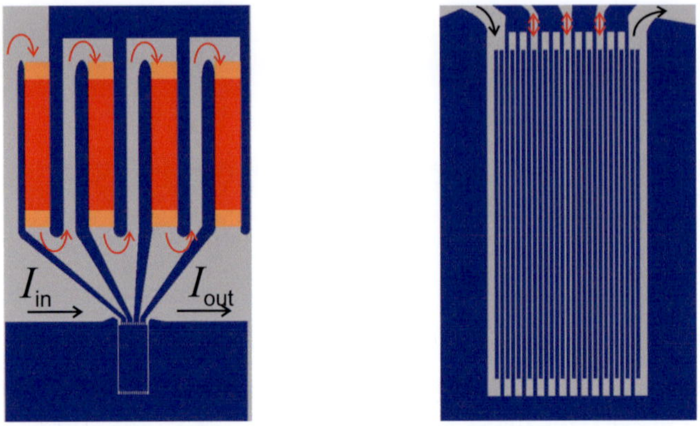

(a) Detector overview

(b) Meandered nanowire

Figure 6.1: The design of a PNR-SNSPD is shown for a 4 pixel detector in the overview (a) and zoomed on the nanowire section (b), the NbN film is shown in gray and the substrate in blue. (a) The resistors are colorized red and the contact pads for the resistors orange. The black arrows show the current flow through the detector and the red arrows the current flow through the parallel resistors. (b) Meander design of the nanowire; the current flow is indicated by arrows, red arrows show the current flow from and to the shunt resistors.

Access to the QD single-photon source, excitation and alignment was done by H. Vural from the University of Stuttgart.

6.1 Detector Design and Fabrication

The design for the PNR-SNSPD used a series SNSPD design as proposed and demonstrated by [101] [102]. The design is explained in detail in section 2.3.9 in chapter 2. The design was used because it allows for the implementation of PNR using amplitude multiplexing and can be used without making changes to an existing SNSPD setup and is operated like a conventional SNSPD. The series-PNR SNSPD design was adapted using the following principles: low fabrication complexity, scalable pixel number, variable absorption length, suitability for top illumination and integration into a tapered WG design (section 4.1) of a QPIC. A local shunting of the nanowire in a spiral detector design is not possible in a single layer, hence the nanowire was arranged in a meander design. To have a scalable number of pixels, the resistors were arranged in a row to allow the addition of further resistors to either side. To enable waveguide integration, the resistors were placed on one side and the nanowire to the other. This allows the placement of the nanowires onto the waveguide and the placement of resistors off the waveguide.

The used detector layout is illustrated for a 4-pixel detector in fig. 6.1. The meandered nanowires are placed at the bottom and the resistors are placed at the top. To prevent detection events in the feed lines used to connect the nanowire to resistors, the width of the feed lines have at least 2.5 times the width of the nanowire. For the resistors, a design value of 25 Ω was chosen, which is on the lower end of previously used values [40]. A small

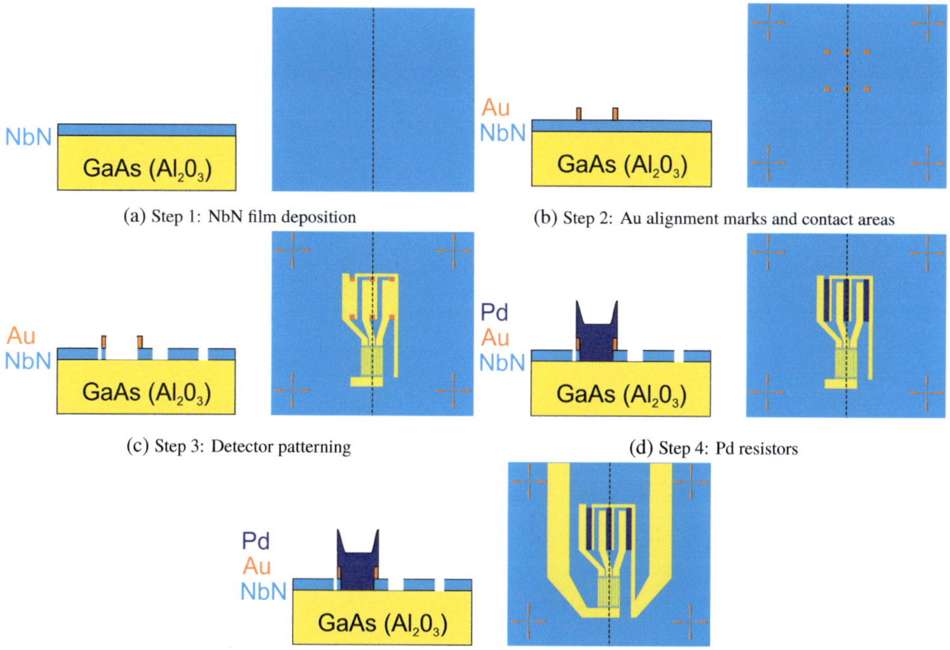

(a) Step 1: NbN film deposition

(b) Step 2: Au alignment marks and contact areas

(c) Step 3: Detector patterning

(d) Step 4: Pd resistors

(e) Step 5: Coplanar waveguide patterning

Figure 6.2: Illustration of the fabrication steps for a PNR-SNSPD.

resistance of the shunt resistors reduces current redistribution out of the detector and ensures a small footprint for the full device. Palladium was chosen as resistor material due to its high corrosion resistance and the high temperature stability of its specific resistance at temperatures <7 K [211]. The resistors are connected to the NbN film by the use of gold contact pads with a width of 5 µm and a length of 2.5 µm to ensure a small contact resistance. The resistors were chosen to a minimal width of 5 µm and a thickness of 30 nm in order to manage a high yield. The µm-width further provides a sufficient contact area and ensures a good adhesion of the Pd to the substrate and a high reproducibility of resistance. A minimal thickness of 30 nm for the resistors is required since the Pd has to bridge the distance of 23 nm from the top of the contact pad to the substrate with a reliable electrical contact. The 30 nm thick Pd film is deposited using DC-Magnetron sputtering at an Ar pressure of 7.1×10^{-3} mbar and a power of 50 W. This translates to a sheet resistance of $R_{\square} = 5.0 \pm 0.3\,\Omega$. Together with the design width and a target value of 25 Ω, the length of resistors has to be 25 µm. The resistors are oriented vertically to save space, so further pixels and resistors can be added to either side. The nanowire is designed in a conventional meander layout (fig. 6.1b). The design width of the nanowire is 100 nm and the gap width between the nanowires is 70 nm.

The fabrication was done in five steps (fig. 6.2a). In a first step, 4.5 nm of NbN was deposited on R-plane sapphire and GaAs(100) substrates. Due to the beneficial effects of an AlN buffer layer seen in chapter 4, a 10 nm thick AlN buffer layer was deposited onto the GaAs prior to the NbN film deposition. In the second fabrication step (fig. 6.2b), alignment marks along

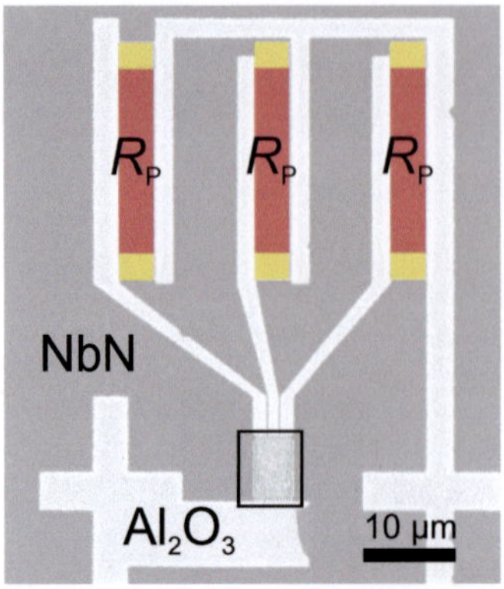

(a) 3-pixel PNR-SNSPD on sapphire

(b) Nanowire of a 3-pixel PNR-SNSPD on sapphire

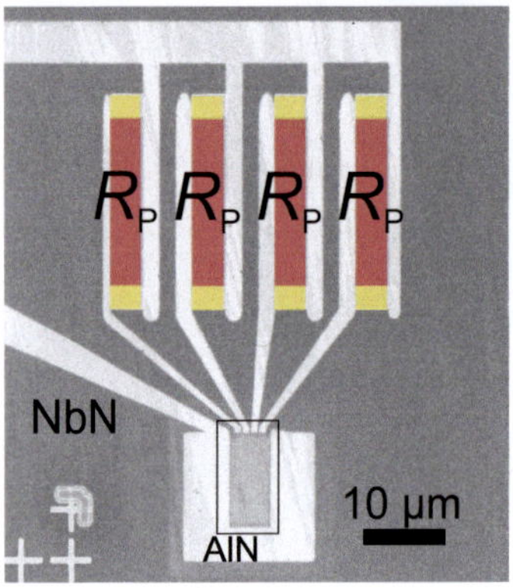

(c) 4-pixel PNR-SNSPD on GaAs ©2019 IEEE [SRS+19]

(d) Nanowire of a 4-pixel PNR-SNSPD on GaAs ©2019 IEEE [SRS+19].

Figure 6.3: Colorized SEM images of PNR-SNSPD on sapphire (a+b) and GaAs (c+d). The darker areas are NbN and the lighter areas the substrate. (a+c) are overview images of the detector including parallel resistors R_P colorized in red and contact areas colorized in yellow. The position of the nanowire that is shown in more detail on the right is marked by a black square. (b+d) are images of the nanowire. The feed lines to connect the resistors are colorized in red, each individual pixel is colorized for clarity.

with contacts and contact areas for the resistors were structured from 15 nm of gold on top of a 3 nm thick Nb adhesion layer using EBL, sputter deposition and lift-off technique. Prior to the Nb/gold deposition, a soft Ar-ion milling step was performed in situ to remove surface contamination and ensure a small contact resistivity and reproducible resistor characteristics. In a third step (fig. 6.2c), the nanowires along with the feed lines to the resistors were patterned using EBL and RIE in a $SF_6 + O_2$ atmosphere. A high stability of both used substrates towards RIE etching in $SF_6 + O_2$ makes it possible to selectively etch the NbN film only. In a fourth step (fig. 6.2d), resistors are structured from palladium using EBL, sputter deposition and lift-off technique. In a final EBL and RIE step (fig. 6.2e), a RF-coplanar feed line is structured on the chip to bias and readout the detectors.

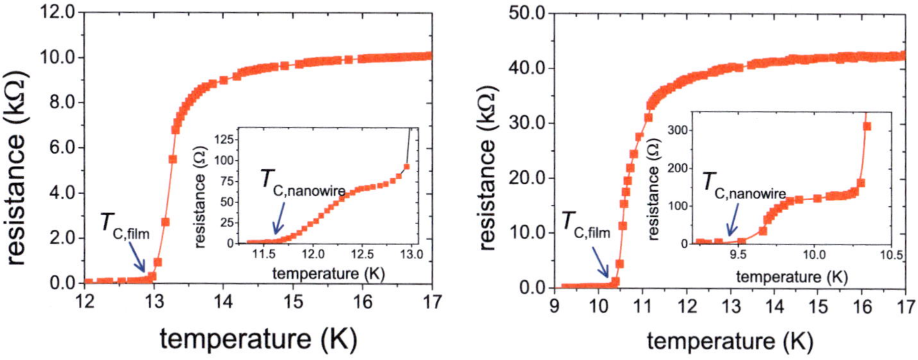

(a) 3-pixel PNR-SNSPD on sapphire

(b) 4-pixel PNR-SNSPD on GaAs ©2019 IEEE [SRS+19]

Figure 6.4: $R(T)$-characteristic of a PNR-SNSPD for each design. The inset is a close up at the foot of their superconducting transition. The transition seen in the inset is caused by the superconducting transition of the nanowire. The resistance in its normal state corresponds to R_P.

A fully patterned PNR-SNSPDs is shown for both substrates in fig. 6.3. The detectors on sapphire (fig. 6.3a, fig. 6.3b) have a wire width of 80 ± 5 nm and a total nanowire length of 144 µm, which is separated by local R_Ps into three pixels with identical wire lengths. For quantum receivers 3-pixels are enough to significantly decrease the error rate during communication [89]. The detector has an active area of 4 µm × 6 µm. The detector on GaAs (fig. 6.3c,fig. 6.3d) has a wire width of 105 ± 5 nm with a total nanowire length of 240 µm. The active area of the detector is 4 µm × 6 µm. The design was stretched to increase the absorption length of the detector for waveguide integration. The used 4-pixel design enables the 4-photon resolution that is required for LOQC.

6.2 Superconducting Properties

The superconducting properties of the detectors were characterized in a 4-probe dipstick setup by measuring the temperature dependence of the resistance $R(T)$ and the current voltage characteristic $I(V)$ at 4.2 K.

The $R(T)$ dependence (fig. 6.4) reveals a two step superconducting transition. Starting at a high temperature, the first step is the superconducting transition of the NbN RF-feed line that is structured out of the same film with a width in the micrometer range. The second transition, seen in the insets, is the superconducting transition of the nanowire. The T_C of the NbN film is 12.8 K on sapphire (fig. 6.4a) and 10.3 K on GaAs (fig. 6.4b). The T_C of the detectors is 11.6 K and 9.4 K, respectively. The small width of the nanowires reduces the T_C by approximately 1 K. The difference in normal state resistance (10 kΩ vs 42 kΩ) for both detectors is mainly caused by a different design of the coplanar readout line which is necessary due to the different RF properties of the substrate and is not connected to film properties. The normal resistance of the second step (see inset) of 70 Ω at 12.6 K for the SNSPD on sapphire

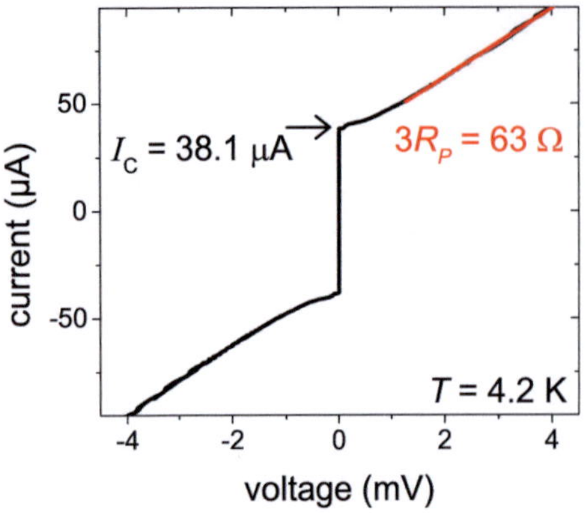

(a) 3-pixel PNR on sapphire

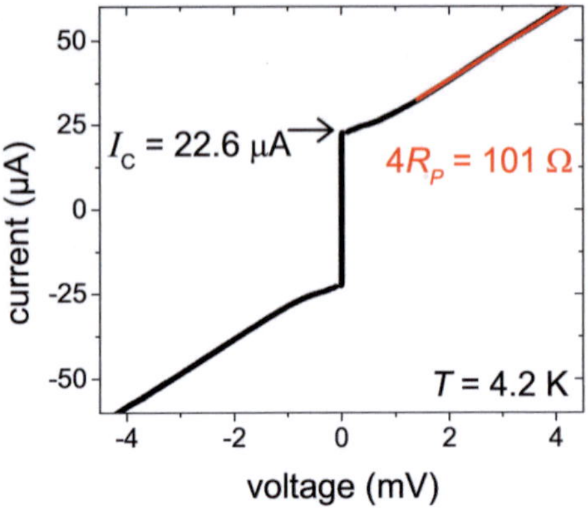

(b) 4-pixel PNR on GaAs ©2019 IEEE [SRS+19]

Figure 6.5: $I(V)$-characteristic of the PNR-SNSPDs. The value of R_p can be extracted from a linear fit of the resistive part of the dependence.

and 120 Ω at 10 K for the SNSPD on GaAs is not determined by the resistance of the nanowire but by R_P since their resistance is four orders of magnitude smaller than the normal resistance of the nanowire.

The current-voltage characteristics for both detectors (fig. 6.5) shows a single-step transition into the resistive state, indicating a high homogeneity of the individual pixels. Furthermore, no hysteresis is visible. This is connected to a redistribution of the current from the nanowire to the shunt resistors, which reduces the thermal load caused by Joule heating in the nanowire and allows for fast recovery into its superconducting state once the current reaches a value lower than the critical current. In comparison to an external shunt resistor, the redistribution of current is faster due to a smaller resistance and capacitance of the connecting feed lines. The I_C of the SNSPD on sapphire is 38.1 μA and of the SNSPD on GaAs 22.6 μA; the higher current of the SNSPD on sapphire is connected to its higher T_C. Out of a linear fit of the ohmic current-voltage dependence in the normal conducting regime, the sum of resistances of all shunt resistors at 4.2 K was extracted. The total resistance of all shunt resistors of the SNSPD on sapphire is 63 Ω, which leads to a value of 21 Ω for each individual resistor. The discrepancy to the targeted 25 Ω can be explained by a slight thickness variation of the sputtered Pd. The extracted 101 Ω for the total shunt resistance of the SNSPD on GaAs fits very well to the design value with a resistance of each resistor of about 25 Ω.

6.3 Experimental Setup for PNR Characterization

For the optical characterization, a free space and a fiber-coupled setup was used. The fiber-coupled dipstick setup is described in more detail in section 4.3 of chapter 4. Since the cooling is done by submerging the end of the dipstick into liquid helium, the operation temperature is 4.2 K. In the free-space setup the SNSPD is excited by photons passing through an optical

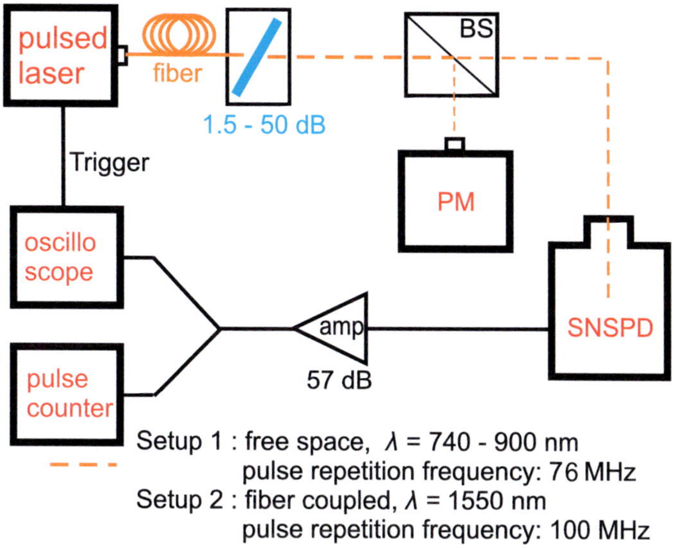

Figure 6.6: Schematic of the utilized measurement setups.

window. A detail description of this setup can be found in section 7.2 of chapter 7. The operation temperature is ≈4 K. In this chapter the free-space setup is further regarded as Setup 1 and the fiber-coupled setup as Setup 2.

The measurement principle of both setups for the characterization of the PNR detector properties is schematically shown in fig. 6.6. Pulsed laser sources with pulsed lengths <5 ps were chosen to investigate the accuracy of the measured results. Lasers have a well-known Poissonian photon statistics and a pulse length significantly shorter than the relaxation time of the detector (few nanoseconds). This ensures a well-known photon statistics for each pulse and allows an evaluation of the accuracy of the measured photon-number (PN) statistic. The used laser source for the Setup 1 is a Ti:Sapphire laser that can be spectrally tuned in the range of 740 to 930 nm and has a pulse length of 3.6 ps at a pulse repetition frequency (PRF) of 76 MHz. Setup 2 operates with a fiber-based fs-laser which operates at a wavelength of 1550 nm and has a pulse length below 150 fs at a repetition rate of 100 MHz. The photons are guided onto a variable attenuator with an adjustable attenuation in the range of 1.5 to 50 dB. After the attenuator, the photons propagate either through free space (Setup 1) or through a fiber (Setup 2) to a 90:10 beam splitter (BS). 90 % of photons are split to a power meter (PM) to control the photon flux and 10 % to the detector. In the cryostat the photon beam is centered onto the SNSPD by the use of a micrometer stage (Setup 1) or micrometer screws (Setup 2) to ensure a homogenous distribution of photons on the detector. The detector is electrically connected to the RF+DC output of a distributed low temperature bias tee. Through the DC path of the bias tee, a bias current is applied to the detector using a battery-driven low noise bias source. Through the RF-path of the bias-tee the detector pulses propagate to a room temperature amplification scheme consisting of three cascaded RF-amplifiers. The full amplifier chain was characterized to have an amplification of 57 dB at 2.8 GHz [132, p. 38].

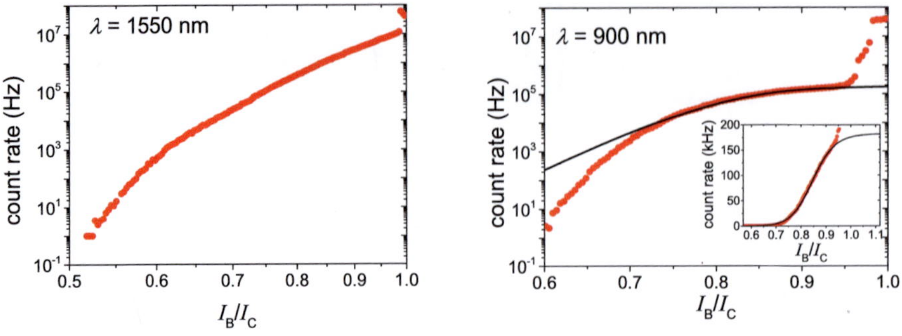

(a) 3-pixel PNR-SNSPD on sapphire

(b) 4-pixel PNR-SNSPD on GaAs ©2019 IEEE [SRS+19]

Figure 6.7: Count rate dependence on the bias current on a single-photon level. The inset of (b) shows the dependence with a linear scaling. The black solid line is a logistic sigmodial fit to the data $< 0.93 I_C$ to extract the saturation level.

The detector pulses were recorded using three different devices: a pulse counting electronic (Stanford Research SR620 Time Interval Counter), a 16 GHz real time oscilloscope and a TCSPC electronic (PicoQuant Hydraharp).

6.4 Optical Characterization

The detectors were studied regarding their single-photon as well their PNR characteristics. Characterization of the detector on sapphire is focused around 1550 nm, which is a wavelength typically used for telecommunication. The detector on GaAs was characterized in a wavelength range from 600 to 1700 nm with a focus on 900 nm, since it is the wavelength of interest for QDs in GaAs photonic integrated circuits.

6.4.1 Bias and Spectral Dependence

The count rate dependence on the bias current was characterized at a mean photon number for each detection event << 1, and a trigger level corresponding to the 1-photon level (see section 6.4.3) of the detector (fig. 6.7). The measurement was performed using a pulse counting electronics. The detector on sapphire was characterized at a wavelength of 1550 nm (fig. 6.7a) and the detector on GaAs at a wavelength of 900 nm (fig. 6.7b). The PNR-SNSPD on GaAs in the exponential scale shows a leveling out of the count rate for bias currents of $0.85 I_C$ which indicates a detector operation close to the deterministic detection regime. From a sigmoid logistic fit to the data $< 0.93 I_C$, a saturation of the IDE of 77 % can be estimated at a bias level of $0.9 I_C$. The sapphire detector does not show a leveling. At $I_B = 0.93 I_C$ the count rate of the detector on GaAs starts to rise until $I_B \approx I_C$ where the pulse counter goes into its counting limit at 36 MHz for both detectors. These additional counts are not caused by photons but are a result of relaxation oscillations. They are caused by the small shunt resistors

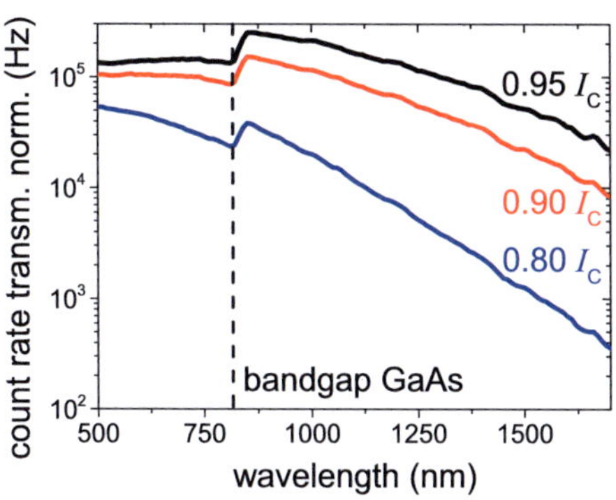

Figure 6.8: Spectral dependence of the detector count rate at different bias currents of a 4-pixel GaAs detector.

separating the detector into smaller pixels with a small kinetic inductance L_{kin}. This leads to a fast recovery of the bias current into the nanowire at a time constant $\tau = L_{kin}/R_P$ [147]. If this time constant is too small, the nanowire cannot fully recover, which in a standard SNSPD causes the detector to stay in a stable resistive state and the detector latches [149]. However, with a small shunt resistor, when the resistivity of the detector rises, the current is immediately pushed back into the shunt resistor, triggering another voltage pulse. The observed oscillations are more significant for the detector on GaAs due to a lower thermal coupling of the film to the substrate fostered by the AlN buffer layer (fig. 3.10d chapter 4). Starting at $0.93 I_C$ the current initially oscillates a few times. A further increase of the bias current increases the number of oscillations until the detector continuously pulses when its critical current is exceeded [212].

The spectral dependence of the detector (fig. 6.8) on GaAs was measured with the pulse counting electronic on a single-photon level using a halogen light source in combination with a monochromator. The shown count rate is normalized for the transmission of the optical setup and the emission of the halogen light source for each measured wavelength. From $\lambda =$ 500 to 815 nm the count rate is more or less stable for $I_B > 0.90 I_C$, which indicates detector operation in the deterministic regime for these wavelengths. An increase in the count rate of 77 % between 810 nm and 840 nm is observed. This is connected to reflected photons at the substrate sample holder interface caused by the GaAs substrate getting transparent for lower energy photons. Afterward, the count rate decreases in the order of one magnitude from 900 to 1700 nm because the detector is operating in the probabilistic detection regime with a decreasing efficiency for lower photon energies. The count rate at $I_B = 0.80 I_C$ decreases in the full investigated spectral range, apart from the characteristic increase at the band gap, and then continues to decrease exponentially. This indicates that the detector operates in the probabilistic regime for $I_B = 0.80 I_C$ over the full investigated wavelength range. The observed count rate dependency of the studied PNR-SNSPDs for excitation on the single-photon level is comparable to a conventional SNSPD.

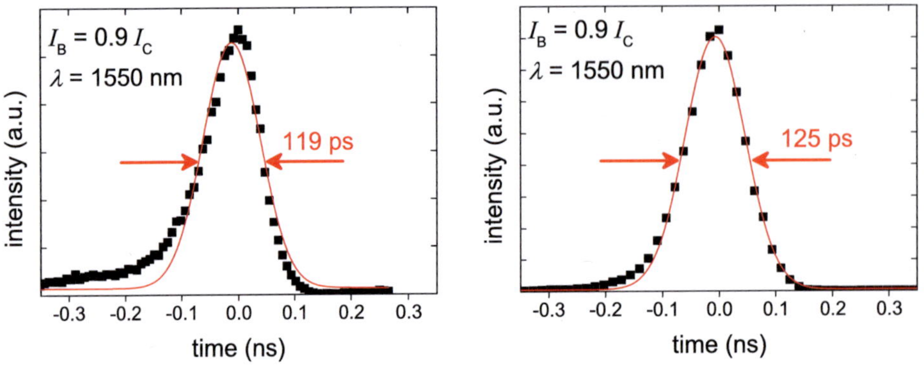

(a) 3-pixel PNR-SNSPD on sapphire

(b) 4-pixel PNR-SNSPD on GaAs

Figure 6.9: Timing histogram for both types of PNR-SNSPDs. The measured data points are shown in black. A Gaussian fit to extract jitter values is shown in a solid red line. The given jitter values are FWHM values.

6.4.2 System Timing Jitter

The system timing jitter is equivalent to the instrumental response function (IRF) of the full measurement system and gives the timing resolution of the full setup. The IRF consists of the timing jitter of the SNSPD, the timing jitter of the photon source, the timing jitter of the optical setup and the timing jitter of the used readout including amplifiers and measurement system (oscilloscope). The system timing jitter was characterized by triggering to a SNSPD pulse at the 1-photon level and recording a histogram of the time deviation of the rising edge of the next trigger pulse at 50 % of the amplitude.

Timing histograms recorded at a wavelength of 1500 nm and a $I_B = 0.90 I_C$ are shown in fig. 6.9. The timing jitter was extracted out the standard deviation σ of a Gaussian fit to the data. Out of σ the FWHM value can be calculated as FWHM $= 2\sqrt{2\ln 2}\,\sigma \approx 2.3548 \cdot \sigma$. The deviation of the measured data to the Gaussian can be attributed to photon dispersion and electronic noise [150] [213]. The FWHM jitter for the PNR-SNSPD on sapphire 119 ps is close to the jitter of the detector on GaAs. The difference can be explained by the length difference, 144 µm compared to 240 µm, and the higher absolute critical current of the PNR-SNSPD on sapphire. A longer length of the device increases the geometrical contribution to the jitter since the transit time of the detector pulse is dependent on the position of the detection spot. A higher critical current causes higher output pulses and a lower influence of noise for the measurement [202]. The measured jitter for the PNR-SNSPDs is larger than the jitter measured for a double spiral in this thesis (section 4.4 in chapter 4).

The system timing jitter for the 4-pixel PNR-SNSPD for different wavelength in the range of λ = 740 to 1550 nm is shown in fig. 6.10. The largest jitter was measured at a wavelength of 740 nm with 220 ps for $0.80 I_C$. For an increasing wavelength from 800 to 900 nm, the jitter decreases by ≈20 %. For λ = 1500 nm, the jitter further decreases from 127 to 108 ps. The jitter decreases with increasing bias current except for an anomaly at 1550 nm in Setup 2 for $I_B = 0.90 I_C$. This decrease is explained by a larger signal-to-noise ratio due to larger

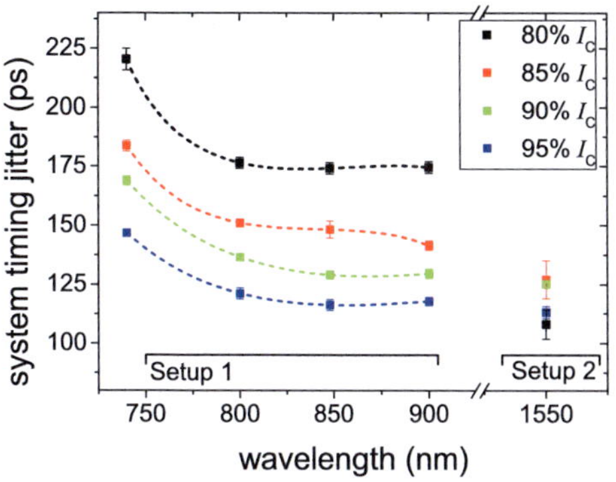

Figure 6.10: FWHM system timing jitter of a 4-pixel SNSPD on GaAs in dependence of the used optical setup and excitation wavelength. The dashed lines are a guide to the eye.

detector pulses at a larger bias current, which reduces the noise contribution to the system jitter [202]. However, the decrease in jitter with increasing wavelength and consequently for lower energy photons is counter-intuitive. The initial hot spot in the nanowire is smaller for a lower photon energy and the time until the full cross-section of the nanowire is normal conducting has a stronger dependence on the position of the initial photon absorption than for photons with higher energy. Smaller jitter values for higher energy photons are usually observed for SNSPDs [150] [42]. The unusual dependence of the measured jitter on the wavelength can be explained by the photon source and the setup used. The significantly lower jitter at λ = 1550 nm is caused by a lower amount of parasitic counts in the fiber-coupled setup (see section 7.3.2). The differences from λ =740 to 900 nm are caused by the used photon source. The mode of the tunable Ti:Sapphire laser was not stable at λ = 740. This led to a jumping of the internal propagating mode and to a not fully deterministic emission of photons. From λ = 800 to 900 nm the mode of the laser was stable and in return the jitter is stable. This demonstrates that for the used setup the contribution of the setup to the system jitter is too large to draw conclusions about a spectral dependence of the detector jitter and that it overshadows the intrinsic jitter of the SNSPD.

6.4.3 Pulse Height and Trigger Levels

To characterize the ability of the SNSPDs to resolve the photon number, the pulse height was investigated. Output pulses measured with the real time oscilloscope are shown in fig. 6.11. A pulse for each amplitude level was recorded at λ = 900 nm for the SNSPD on GaAs at $I_b = 0.9 I_C$. These pulses were measured at a high photon flux with a large mean number of photons per pulse (μ) ($\mu >> 1$). The blue pulse represents the detection of one photon, the green pulse of two, the red pulse of three and the black pulse of four photons. The $1/e$ decay time ($\tau_{1/e}$) of the measured pulses is approximately 4.4 ns.

The height of pulses was investigated by the use of different voltage thresholds for the pulse counter. A pulse is counted at its rising edge as long as its amplitude is higher than the

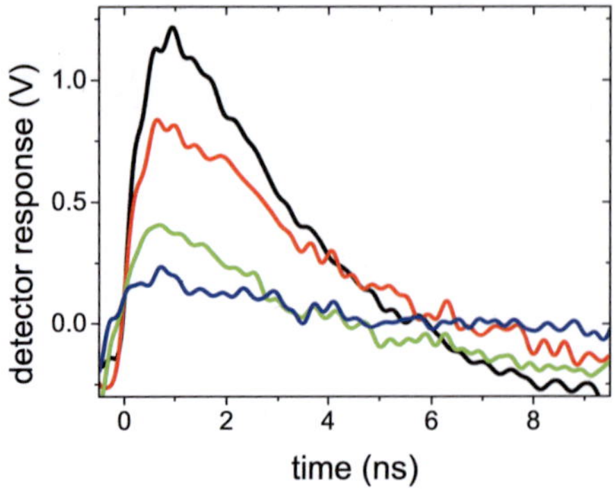

Figure 6.11: Output pulses of a 4-pixel PNR-SNSPD on GaAs for each individual amplitude level. ©2019 IEEE [SRS+19]

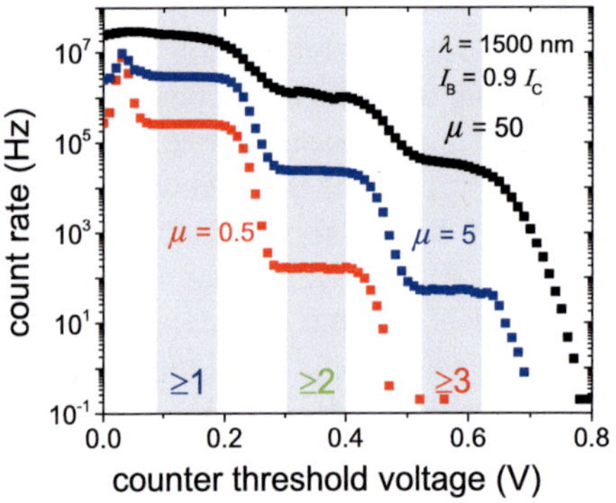

(a) 3-pixel PNR-SNSPD on sapphire

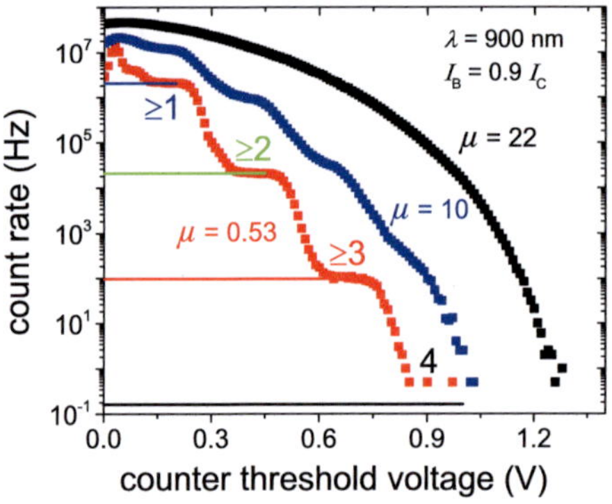

(b) 4-pixel PNR-SNSPD on GaAs ©2019 IEEE [SRS+19]

Figure 6.12: Count-rate dependence on the threshold voltage of the pulse counter at different mean photon numbers per pulse μ for a 3-pixel PNR-SNSPD on sapphire and a 4-pixel PNR-SNSPD on GaAs using a pulsed laser source. The numbers in the graph correspond to the individual photon levels.

threshold voltage level. The threshold level dependence is shown at $\lambda = 1550$ nm for the detector on sapphire (fig. 6.12a) and at $\lambda = 900$ nm on GaAs (fig. 6.12b). At first, only the red data points for $\mu \approx 0.5$ are considered. In fig. 6.12a, after a first initial spike at ≈50 mV three equidistant ($\Delta V = 200$ mV) flat regions marked in gray are visible. The spike is caused by the readout noise. The flat regions on the other hand are caused by the rising flank of the detector pulses. The presence of three flat regions proves that the detector is able to discriminate between three PN levels. Starting from the left, the first flat region marks the detection of one or more, the second flat region of two or more and the third flat region of three photons. The photon count rate for each PN can be conveniently measured by triggering at the corresponding flat region and subtracting the count rate of the next higher amplitude region. For the 4-pixel PNR-SNSPD on GaAs, four equidistant ($\Delta V = 250$ mV) flat regions are visible, proving the discrimination of four PN levels. The smooth transition between the discrete flat levels is caused by the amplitude jitter. For an increasing μ, the overall count rate increases and the amplitude jitter increases. The amplitude levels start to overlap until no discrete flat regions are visible (black curve for $\mu = 22$ in fig. 6.12b). The increasing amplitude jitter is caused by a shift of the ground potential based on the detection of the previous pulse. The strength of this shift is hereby proportional to the detected PN of the previous pulse (fig. 6.11). After a pulse on the 1-photon level (blue), the ground potential reaches 0 mV after 9 ns. For a four photon detector pulse the ground potential is at ≈ -0.5 V after 9 ns. For a pulse repetition frequency (PRF) of the laser of 100 MHz, the next laser pulse is emitted after 10 ns which is too fast at a large μ and the readout cannot fully relax. For a small μ, not all laser pulses trigger a detection event and most detection events are on a small PN-level so the readout line has enough time to relax. However, when the flux is increased and more detection events occur, in addition, to more large PN-detection events, the amplitudes start to overlap. In direct comparison, a larger amplitude jitter is seen for the PNR-SNSPD on GaAs. This can be explained by taking into account the detection efficiency and the pulse height. The detection efficiency of the SNSPD on GaAs at a wavelength of 900 nm is higher than the detection efficiency of the SNSPD on sapphire at 1550 nm, so overall more laser pulses and larger PNs are detected on the GaAs SNSPD for the same μ. Furthermore, the average amplitude of detector pulses is smaller for the SNSPD on sapphire since it only has three pixels and a smaller R_P.

The SNSPD on GaAs was further tested for constant light sources (fig. 6.13): for a LED at $\lambda = 385$ nm (fig. 6.13a), a 652 nm continuous wave (CW) laser diode (fig. 6.13b) and a halogen light source in combination with a monochromator (fig. 6.13c) at $\lambda = 900$ nm. For all light sources, three individual steps can be seen. This proves that the PNR-SNSPD is able to discriminate between at least three PN levels for the shown three constant light sources. For a voltage below 150 mV, a flat region is visible, which is not an actual flat region caused by the rising edge of detector pulses but is caused by the count rate limit of the used counting electronics. One can see that, since the bias level influences the detection efficiency, a higher count rate for higher relative bias currents is visible. Due to the number of counts, the amplitude jitter is increased. A four-photon step is not visible because of a to small power level of the light sources. Since the investigated sources shown in fig. 6.13 are constant light sources, photons reach the detector separated in time and not in a well-defined time slot. Therefore, the amount of photons that reach the detector at a time is reduced. In addition, the amplitude levels are smeared out since new detection events occur while the previous detector pulse has not fully relaxed. Hence, a pulsed source is needed for a defined photon statistics on the detector.

In summary, the fabricated PNR-SNSPDs show amplitude multiplexing with a number of amplitude levels proportional to the number of pixels. This shows that all detectors are able to resolve a PN equal to their number of pixels. This successfully demonstrates PNR with amplitude multiplexing. At a low photon flux for a pulsed light source, the PN levels are well separated and a pulse counting electronic adjusted to the corresponding voltage level is enough to resolve the detector output. At a high photon flux the amplitude jitter gets too large and the PN levels cannot be easily discriminated by a simple pulse counter. In conclusion, photon-number resolution is demonstrated for all tested light sources.

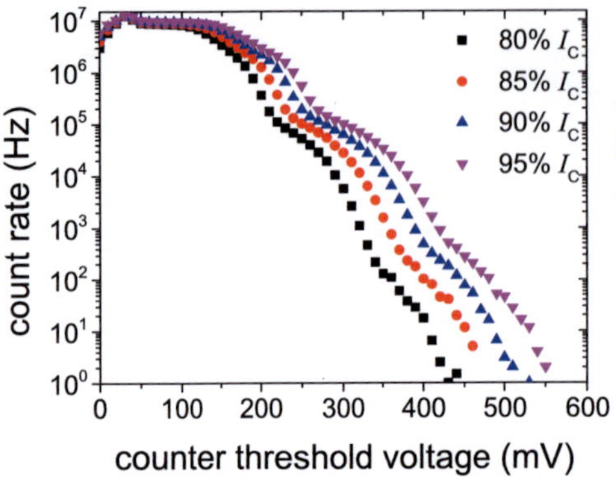

(a) 385 nm LED light source at ≈75 pW

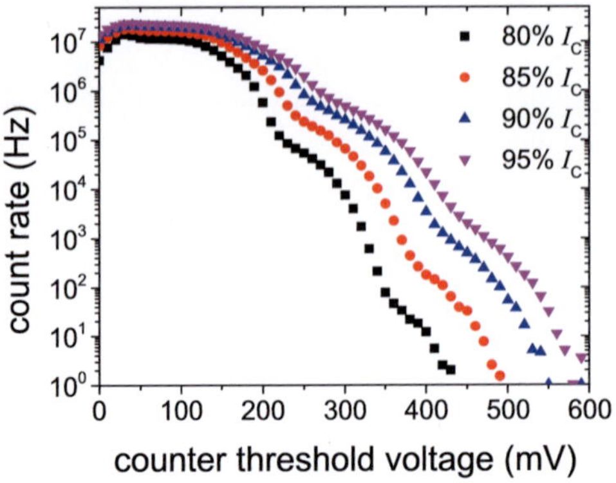

(b) 652 nm CW laser light source at ≈75 pW

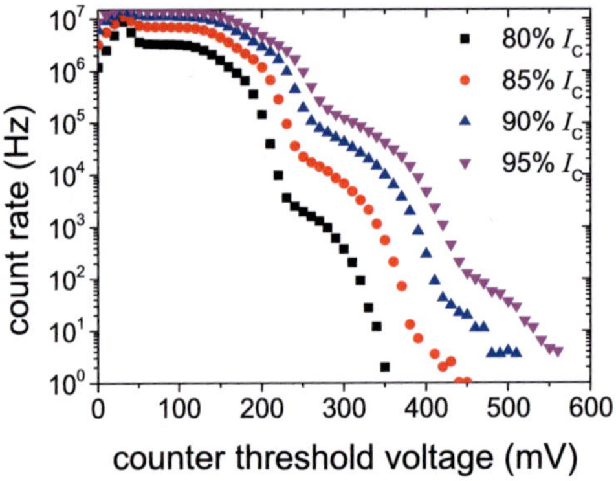

(c) 900 nm thermal light source at ≈18 pW

Figure 6.13: Count-rate dependence on the threshold voltage of the pulse counter for different light sources at different bias levels for a 4-pixel GaAs PNR-SNSPD.

6.4.4 Photon Statistics

The accuracy of the detector statistics to resolve the PN distribution of a laser light source was characterized by measuring the statistical distribution of pulse amplitudes for a given μ using a real time oscilloscope by recording at least 200 000 detector pulses. The trigger output of the corresponding laser source was used as a time reference. The data for a $\lambda = 740$, 800 and 900 nm was obtained in Setup 1, and for $\lambda = 1550$ nm in Setup 2. The detectors were adjusted to $I_B = 0.9 I_C$.

A heat map of 250 000 detector pulses of a 3-pixel PNR-SNSPD on sapphire recorded at $\mu = 146$ is shown in fig. 6.14. The three amplitude levels are clearly visible. Due to the large μ, the 0-photon level is only slightly visible. The white line gives the position where the pulse amplitudes were recorded and the corresponding histogram is shown in cyan on the left.

Gaussian fits were used to extract the PN levels out of the overlapping pulse amplitudes at a large μ. A histogram of the amplitude distribution normalized to one at $\lambda = 1550$ nm for

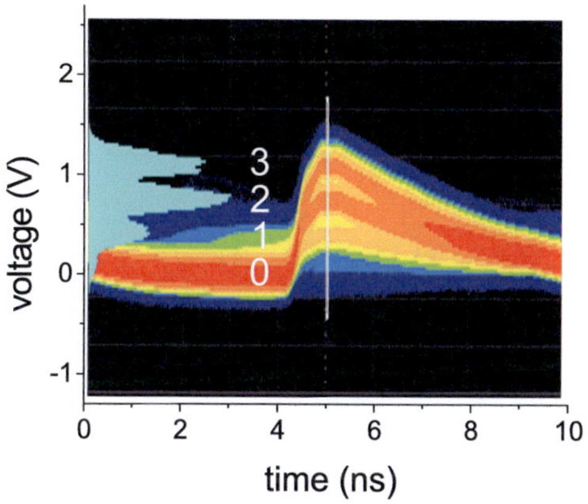

Figure 6.14: Heatmap of 250 000 pulses for a 3-pixel PNR-SNSPD on sapphire, the white line is the position where the pulse amplitudes were recorded. The number indicates the corresponding PN level of the amplitude. The corresponding histogram is shown in cyan on the left.

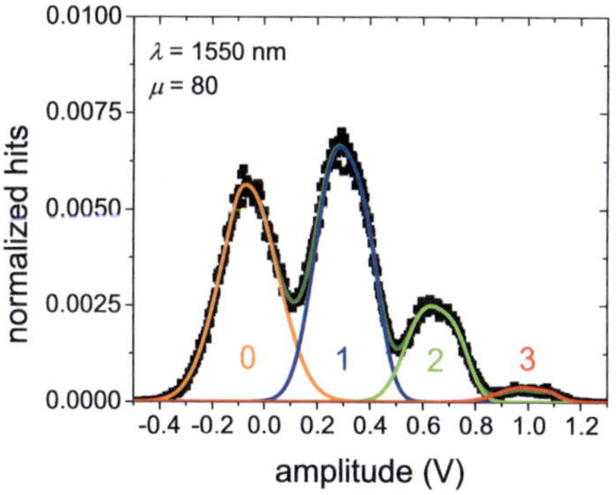

(a) 3-pixel PNR-SNSPD on sapphire

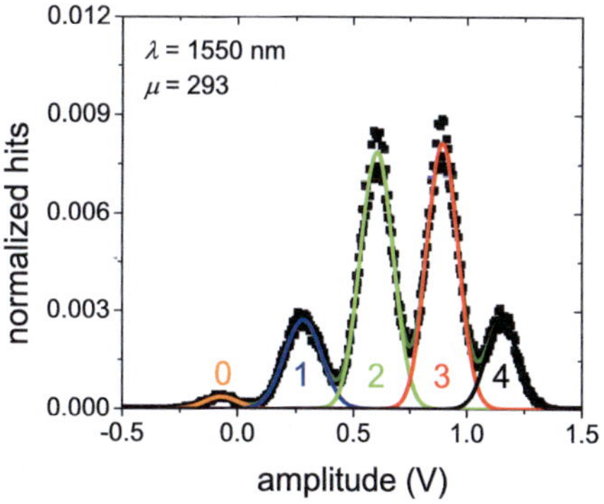

(b) 4-pixel PNR-SNSPD on GaAs

Figure 6.15: Histograms of the amplitude distribution at $\lambda = 1550\,\text{nm}$. The solid lines are Gaussian fits to the individual photon number levels and the sum of all Gaussian fits (dark green).

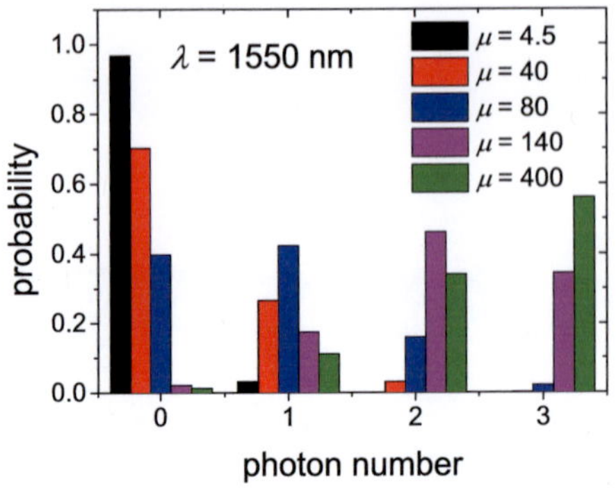

Figure 6.16: Photon-number probability distributions of a pulsed laser at different μ measured with a 3-pixel PNR-SNSPD on sapphire.

a 3-pixel detector on sapphire is shown in fig. 6.15a and for a 4-pixel detector on GaAs in fig. 6.15b. Each histogram was fitted with a sum of Gaussian distributions (dark green) with a Gaussian for each amplitude level of the detector including the 0-photon level (no detected photons for a laser pulse (orange)). During the characterization of the SNSPD on sapphire (fig. 6.15a) two peaks were present at each amplitude level due to reflections in the readout and a sum of two Gaussian distributions was necessary to correctly extract the pulse amplitude distribution. The sum of all Gaussian-functions (dark green line) fits well to the underlying data (black dots). Due to the normalization, the probability of a PN is given by the area of the corresponding Gaussian.

The extracted PN probability distribution in dependence on μ of a 3-pixel PNR-SNSPD on sapphire is seen in fig. 6.16 for λ = 1550 nm. The PN-distribution shows a clear dependence on the incoming mean number of photons pro pulse. At a small μ mainly 0 or 1 photons are detected for each pulse. With increasing flux the PN probability distribution shifts to larger photon numbers since the detection of higher PN-levels gets more probable.

The same can be seen for a 4-pixel PNR-SNSPD on GaAs (fig. 6.17). The probability distribution shows a shift to larger PN levels for a higher photon flux at all investigated wavelengths. The shift of the distributions to higher photon numbers for an increasing μ is stronger for shorter wavelengths. This is expected since the detector has a higher DE at higher photon energies (section 6.4.1) which significantly increases the probability of correctly detecting a multi-photon event.
To evaluate the accuracy of the measured photon probability distributions, the measured PN-statistics of the 3-pixel PNR-SNSPD on sapphire at 1550 nm and of the 4-pixel PNR-SNSPD on GaAs at λ = 900 nm were fitted with the theoretical PN-statistics $P_\eta^N(n|\mu)$ (eq. (2.13) in section 2.3.9 in chapter 2) expected for a laser using the detection efficiency η as a free fit parameter. N is the number of pixels and n the detected number of photons. For a small μ, the fit closely represents the experimental data for both detectors (fig. 6.18). A direct comparison of the fit to the measured statistics for the 3-pixel PNR-SNSPD on sapphire for μ = 0.4 is shown in fig. 6.19 and shows that the fit is in good agreement with the experimental data

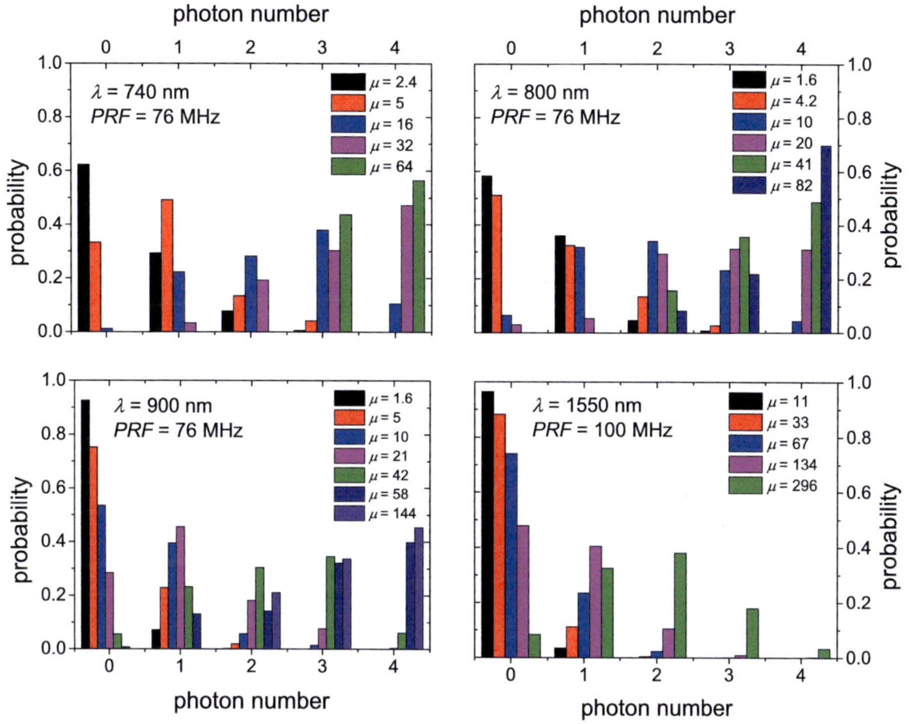

Figure 6.17: Distribution of photon-number levels for different wavelengths of a pulsed laser measured with a 4-pixel PNR-SNSPD on GaAs.

at a small μ. The probability distribution is shown in logarithmic scale to visualize the full statistics. The lack of a recorded detection probability for the experimental 3-photon level is connected to the number of recorded pulses. For the shown figure, the statistics was done over 300 000 pulses, so the probability of measuring a 3-photon pulse in the observed time frame is less than 0.2 %.

For an increase of μ the fit starts to deviate from the experimental data. The extracted detection efficiency (η) values (black) out of the fit of $P_\eta^N(n|\mu)$ for each μ are shown in fig. 6.20 along with the absolute deviation (red) of the fit from the experimental data for each value of μ. The absolute deviation is the sum of the differences between the experimental data and the fit for each PN.

For the 3-pixel PNR-SNSPD on sapphire (fig. 6.20a) for a $\mu \leq 60$, the absolute deviation of the fit to the experimental data is <1 %. In addition, the extracted η agrees very well with the DE measured on a single-photon level with DE = 8.5 % (blue line). At larger values of μ the absolute deviation is significantly larger than 1 % which shows that the measured PN does not fully represents the Poissonian statistics of the light source. The η slightly rises with the increase in μ. The rise until $\mu = 160$ is connected to the used battery-driven bias source.

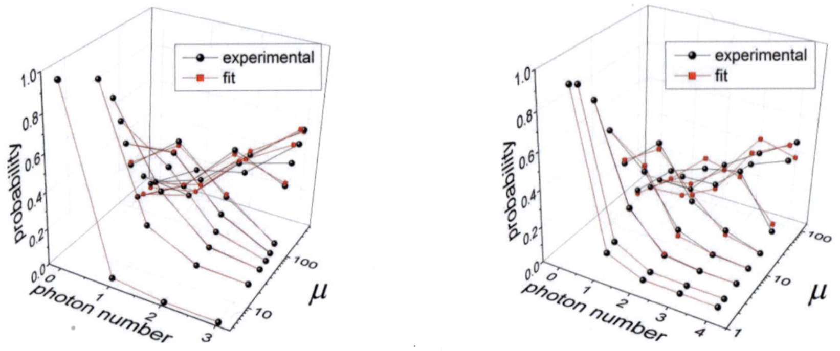

(a) 3-pixel PNR-SNSPD on sapphire

(b) 4-pixel PNR-SNSPD on GaAs ©2019 IEEE [SRS+19]

Figure 6.18: Comparison of the experimental PN distributions at different μ with the fit of the theoretical distribution. (a) PN distribution for the 3-pixel PNR-SNSPD on sapphire at λ = 1550 nm. (b) PN distribution for the 4-pixel PNR-SNSPD on GaAs at λ = 900 nm.

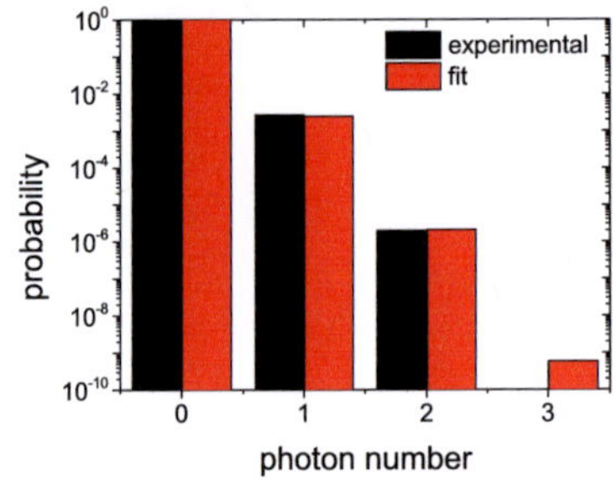

Figure 6.19: Experimental data and fit for the probability distribution for μ = 0.4 at 1550 nm for a 3-pixel PNR-SNSPD on sapphire.

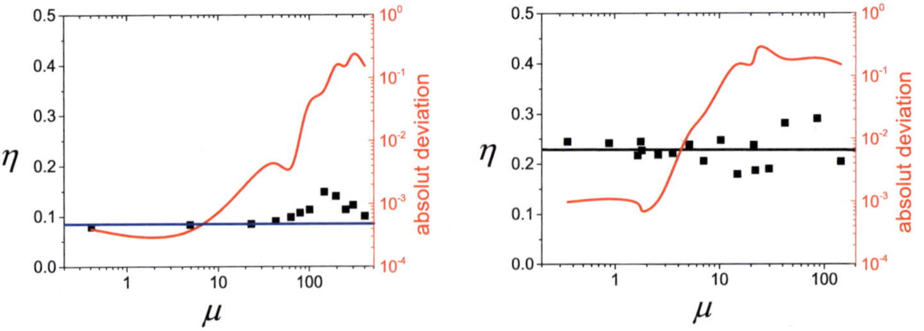

(a) 3-pixel PNR-SNSPD on sapphire

(b) 4-pixel PNR-SNSPD on GaAs ©2019 IEEE [SRS+19]

Figure 6.20: For both detectors, the detection efficiency η extracted out of the theoretical fit is shown for (a) λ = 1550 nm and (b) λ = 900 nm. The blue line in (a) is the DE calculated at the single-photon level and the black line in (b) is a linear fit to η. In both, the absolute deviation of the the theoretical distribution $P_{\eta}^{N}(n|\mu)$ (eq. (2.13)) to the experimental measured data is shown in red.

The histograms were measured in sequence starting from the lowest μ to the highest μ. Each PN is measured statistically and the number of triggered measurement events, that the used oscilloscope can measure per second, is limited. Accordingly, one data point was measured over the time of approximately one hour. A low battery level led to a shift of the reference voltage which led to a slowly shifting false reading of the applied bias level. To always operate at the same bias condition, the output was adjusted to the same bias level for each point, which led to a slight deviation from the previous bias condition due to the shifted reference. Since the detector was not operating in the deterministic region, a small shift in the bias level results in a significant change in the DE (see fig. 6.7a). The seen η difference can be explained by a 2 % bias current difference. For $\mu > 160$, the battery was replaced and the measured scatter of η is connected to the higher absolute deviation of the fit from the measured data. For a 4-pixel PNR-SNSPD on GaAs and $\mu < 5$, the absolute deviation is less than 1 % and the η shows only a small scatter. For higher values of μ, the scatter in η is significant and the fit shows a deviation from the experimental data in the order of 20 %. A linear fit to the full data set reveals a flat average η of 22.7 ± 0.3 % (fig. 6.20b). This η is in good agreement with the distribution of flat regions seen in dependence of the used pulse-counter threshold voltage in the previous section (fig. 6.12b). The stable η indicates that the count rate of the detector is sufficient to resolve the PN distribution of a light source at a PRF of 76 MHz for the investigated range of μ from 0.35 to 144 in contrast to the used readout. The good agreement of the fit with the measured data for $\mu < 60$ for the SNSPD on sapphire and for $\mu < 5$ for the SNSPD on GaAs, respectively, demonstrates that the measured statistics closely represents the statistics of the light source for small values of μ and furthermore, shows a high homogeneity of the detector pixels. The large scatter of η and the deviation of the fit from the experimental distribution can be assigned to the readout. The described merging of amplitude levels for large μ (section 6.4.3) makes the fitting of the measured amplitude distributions with Gaussian-fits inaccurate for the extraction of the PN distribution at large μ. This behavior also explains the larger measured dynamic range of the PNR-SNSPD on sapphire. Since the detection efficiency at the respective used wavelength is lower for the sapphire detector, fewer photons are detected overall and the PN remains lower for the same values of μ. In addition, the SNSPD on sapphire has one pixel less and has a lower resistance of the parallel resistors, which results in a lower charge transfer. Consequently, a higher dynamic range for statistical measurements with a series PNR-SNSPDs can be achieved by decreasing its DE. A lower DE can be obtained by lowering its bias current, at the cost of sensitivity and fidelity for a single-shot measurement. Furthermore, an optimization of the readout for a reduced shift in the ground potential would lead to a higher dynamic range.

In conclusion, the investigated series PNR-SNSPDs are able to accurately resolve the PN out of a pulsed Poissonian light source with a photon number per pulse of at least $\mu = 60$ for the PNR-SNSPD on sapphire and $\mu \leq 5$ for the PNR-SNSPD on GaAs respectively. A high accuracy for a $\mu \leq 5$ for the tested 4-pixel PNR-SNSPD on GaAs in combination with the used setup is sufficient for LOQC [13].

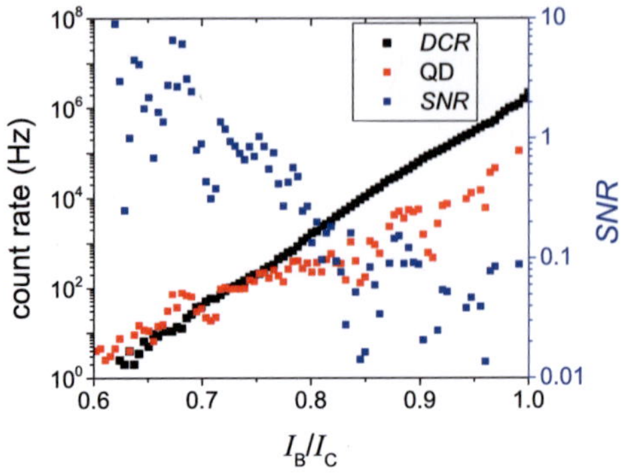

Figure 6.21: Count rate on the detector for QD photons (red) and dark counts (black). The resulting SNR is shown in blue.

6.5 Discriminating a Poissonian from Sub-Poissonian Light Source

To evaluate the benefit of a series PNR-SNSPD for a photonic integrated-circuit, the response of a 4-pixel PNR-SNSPD on GaAs to laser emission (Poissonian) is compared to the response to QD emission (sub-Poissonian). The experiment was performed in Setup 1. The laser source used was the previously used pulsed Ti:Sapphire laser. The In(Ga)As/GaAs quantum dot was excited by a pulsed resonant laser in a separate flow cryostat. Details on the QD and its excitation can be found in [214]. The QD is a high purity single-photon source with a $g^{(2)}(0) = 0.013 \pm 0.007$, a decay time of 0.53 ns and an emission wavelength of 854 nm [214].

In order to find the optimal I_B for the SNSPD, the photon count rate for the QD emitted photons was recorded and compared to the DCR. One can directly see, that the rate of dark counts at a high bias current is higher than the count rate for QD emitted photons. The signal-to-noise ratio (SNR) improves for a small I_B/I_C. However, a low bias current is not optimal, since the DE is low, which leads to a significantly smaller count rate and, even more important, to a decreases in detection probability for multi-photon events. As a compromise between SNR and DE, a $I_B = 0.80 I_C$ was chosen. The laser emission was adjusted to $\lambda = 854$ nm and the photon flux to a count rate of $\approx$100 Hz on the SNSPD to have a comparable wavelength and photon flux as the QD emission. To remove dark counts from the measurement, a Picoquant HydraHarp TCSPC electronics was used. TCSPC stands for time correlated single-photon counting and is a general term for time correlated measurements using single photons. The time tagging of incoming detector pulses allows the separation of dark counts from the pulsed detector signal, since the dark counts are independent on time and can be seen as a base noise level. The trigger signal of the corresponding laser, either the Ti:Sapphire laser when used as photon source or the excitation laser for the QD, were used as a time reference. The detector response was recorded at the 1- and 2-photon level by triggering to the corresponding voltage levels simultaneously, using different trigger levels at separate input channels. Since $\mu << 1$, the amplitude levels of the detector were well separated.

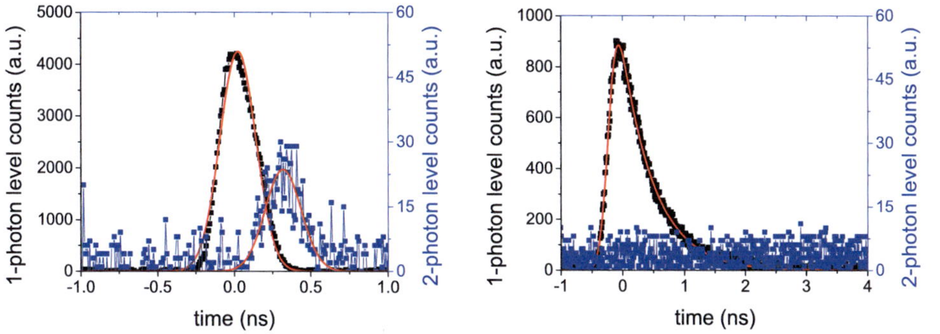

(a) Pulsed laser

(b) Quantum dot

Figure 6.22: 2-photon level TCSPC measurement for a pulsed laser (a) and a resonantly excited quantum dot single-photon source (b). The 1-photon level is displayed in black and the 2-photon level in blue. The red solid lines are fits to the data: (a) Gaussian fits, (b) Exponentially modified Gaussian ©2019 IEEE [SRS+19]

The recorded histograms are shown in fig. 6.22 after subtraction of the base noise level. The detector's response to the laser is shown in fig. 6.22a. Photons are detected on the 1- (black) and 2-photon level (blue). The 2 photon level (blue) is recorded with a delay of 232 ps due to a slightly longer cable. The timing distribution on both photon levels corresponds to the IRF of the setup when using the TCSPC-electronics. Out of a Gaussian fit (red line), a standard deviation of σ = 98 ps is retrieved for both photon levels. The detector's response to the QD emission (fig. 6.22b) shows a clear signal on the 1-photon level and only noise on the 2-photon level. The absence of a 2-photon level for the QD photons in direct comparison with the laser at a similar flux, clearly shows that the detector resolves the sub-Poissonian character of emitted photons. The timing distribution was fitted with an exponentially modified Gaussian distribution (EMG) (red line). The EMG resembles the Gaussian distribution of the IRF convoluted with a mono-exponential probability density function to model the mono-exponential decay of a QD. Out of the fit an IRF of $\sigma = 114 \pm 1$ ps was extracted for optical excitation with the QD and a decay time for the QD of 525 ± 3 ps. This decay time is in good agreement with the previously measured decay time of 0.53 ns [214]. This shows, that the PNR-SNSPD presented in this chapter is able to successfully discriminate a light source with a sub-Poissonian from a light source with a Poissonian statistic.

6.6 Summary

In this chapter, a design of photon-number resolving SNSPDs suitable for waveguide integration was demonstrated. The design is based on the series photon-number approach presented by Jahanmirinejad in 2012 [101]. Detectors were fabricated in two slightly different designs on sapphire and GaAs substrate. The system timing jitter for detectors on both substrates was characterized to be in the range of 130 ps. The first series PNR-SNSPD on sapphire was demonstrated with a photon-number resolution of 3. The detector was shown to accurately represent the PN statistics of a 1550 nm fs-laser with a pulse rate frequency of 100 MHz for a mean photon-number below 60 photons/pulse with a detection efficiency of 8.5 % at 4.2 K. A photon-number resolution of 4 was demonstrated for a NbN SNSPD on GaAs optimized for integration into a waveguide. The characterization was performed in a free-space setup at 4 K for wavelengths from 740 to 900 nm and in a fiber-coupled setup at 4.2 K. At a wavelength of 900 nm and $I_B = 0.9 I_C$ the intrinsic detection efficiency of the detector was estimated to 77 %. A 4-photon resolution was successfully shown for a pulsed and cw-laser, an UV-LED and a thermal halogen light source. The detector was shown to accurately represent the photon-number statistics of a ps-laser, at a wavelength of 900 nm, a pulse rate frequency of 76 MHz for a mean photon rate below 5 photons/pulse, at a detection efficiency of 22.7 ± 3.0 %. It was found that the used readout, limits the dynamic range and jitter of the detectors. For a characterization of the physical limits of the jitter and the dynamic range of the detectors, the readout needs optimization. The response of the detector on GaAs to a high-quality GaAs-based QD single-photon source was tested and the ability to discriminate the statistics of photons of a sub-Poissonian from a Poissonian was demonstrated. This shows the practical benefit of PNR-SNSPDs for the field of quantum information technology. With the longer absorption length on a waveguide, together with an increased intrinsic detection efficiency using the previously discussed variable thickness approach, a detection efficient close to unity is in reach for a waveguide integrated PNR-SNSPD with the shown film quality.

7 Monolithic On-Chip Integration of a Full Quantum Photonic Circuit

In this chapter, the first full integration of a QPIC on a monolithic platform is demonstrated using GaAs as a substrate. The used approach utilizes InGaAs/GaAs QDs which are embedded into single-mode GaAs/AlGaAs-based rib waveguides. A double-spiral type NbN-SNSPD is used as a detector in combination with an AlN buffer layer (see chapter 4). The highest degree of indistinguishability for emitted photons is achieved by excitation of the QDs using an excitation laser in resonance to the QDs energy band [16]. Furthermore, resonant excitation allows the selection of a single dot out of a large number of randomly assembled dots. However, the excitation laser causes a significant amount of stray light on-chip seen by the integrated SNSPD. A spectral filtering of stray light is not possible beacuse for resonant excitation, photons used for excitation and photons emitted by the excited QD have the same wavelength. In the first part of this chapter, the amount and predominant origins of stray lights are investigated and methods for suppression and increase of the signal-to-noise-ratios are developed. In the second part, a fully integrated device is introduced and characterized followed by the first successful fully on-chip Hanbury-Brown and Twiss experiment [107] on a monolithic platform.

Parts of the work presented in this chapter have been published in Nano Letters under the title: *"Fully On-Chip Single-Photon Hanbury-Brown and Twiss Experiment on a Monolithic Semiconductor-Superconductor Platform"* [SSR$^+$18]. The substrates with integrated QDs were grown by S. Hepp from the University of Stuttgart. The used waveguides were designed, simulated and etched by U. Rengstl from the University of Stuttgart in the scope of a PhD thesis [110]. The electron beam lithography of waveguides was performed by T. Reindl of the Max Planck Institute for Solid State Research in Stuttgart. Off-chip optical measurements, the adjustments of the free space optical setup and the excitation of QDs were done by M. Schwartz in the scope of a PhD thesis [124] with support from T. Herzog [215], J. Binz [216] and F. Hornung [217], all from the University of Stuttgart. The author of this thesis made the chip designs with support from U. Rengstl, the fabrication of detectors and covers and all fiber-coupled optical measurements. On-chip optical measurements were performed together with M. Schwartz with support by T. Herzog [215], J. Binz [216] and F. Hornung [217]. A description of the experiment with more focus on QDs and QD excitation can be found in the PhD thesis of M. Schwartz [124].

(a) Substrate

(b) WG Layer Structure

Figure 7.1: a) Sketch of the layer structure of the grown substrate. b) Sketch of the cross section of a single-mode rib WG.

7.1 Design and Fabrication of Basic Elements

7.1.1 Basic Design

The basic design of the implemented photonic integrated circuit consists of a quantum dot InGaAs/GaAs single-photon source, a single-mode GaAs waveguide structure and a SNSPD. The basic layer structure of the chip consists of a 350 µm thick GaAs (100) substrate with a 2 µm thick $Al_{0.4}Ga_{0.6}As$ cladding layer followed by a 370 nm thick GaAs core layer. In dependence of the experiment, either a dual or single-side polished GaAs (100) substrate was used. The $Al_{0.4}Ga_{0.6}As$ layer has a lower refractive index compared to the GaAs core layer and acts as a lower confinement layer, preventing photons from escaping to the bulk substrate. Self-assembled InGaAs quantum dots are integrated 135 nm above the upper $Al_{0.4}Ga_{0.6}As$/GaAs interface to match the expected maximum of the distribution of the evanescent field in the final waveguide structure. The quantum dots emit at a wavelength of approximately 900 nm. An approximately 580 nm wide, 320 nm high, 2 mm long single-mode rib waveguide is structured out of the core layer with a 50 nm high slab (fig. 7.1b). In one direction, the WG is tapered in a linear shape towards the detector over a distance of 250 µm to a final width of 10 µm to embed the entire SNSPD. In the other direction, the sample is cleaved perpendicular to the WG to enable free space coupling. Cleaving is performed by scoring the sample on the surface in a controlled manner and afterwards breaking along the scored surface. During this process single crystalline GaAs substrate breaks along its crystalline axis and forms an atomic flat breaking surface. For the detector, the previously introduced (chapter 4) double-spiral SNSSPD design was used. The SNSPD is connected to a coplanar RF waveguide to electrically contact the SNSPD for biasing and readout of the detector pulses. The design of the detector and RF readout line is described in detail in section 4.1 of chapter 4. For the final experiments, an Al/AlN bilayer was added on top to shield the detector from stray light.

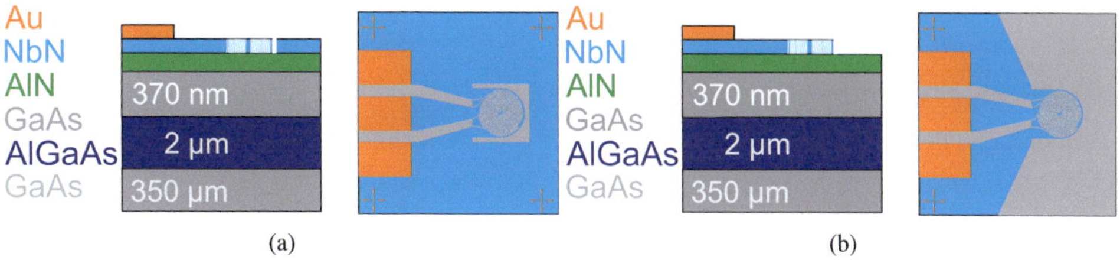

Figure 7.2: a) Detector after the fabrication process described in section 4.2 of chapter 4. b) Detector after NbN removal.

7.1.2 Substrate Growth and Embedding of Quantum Dots

The basic layer structure of the substrate as seen in figure 7.1a was grown at the University of Stuttgart by Stefan Hepp using a commercial Aixtron 200 Metal-Organic Vapor-Phase Epitaxy (MOVPE) system with standard precursor gases: trimethylgallium (TMGa), trimethylindium (TMIn), trimethylaluminum (TMAl) and arsine (AsH_3). The growth was performed at a pressure of approximately 100 mbar on GaAs (100) substrates oriented 6° toward the (111) crystal orientation. The GaAs substrate is heated to a temperature of 710 °C to desorb surface oxides. On the bare surface a 100 nm thick GaAs buffer layer is grown. This buffer layer is followed by a 2 µm thick $Al_{0.4}Ga_{0.6}As$ layer, which acts as a lower cladding layer for the later etched waveguides and prevents carrier diffusion towards the substrate during optical measurements due to its higher bandgap compared to GaAs. On top of the $Al_{0.4}Ga_{0.6}As$ a 135 nm thick GaAs layer was grown, followed by 2-4 monolayers of InAs using the Stranski-Krastanov growth mode [122] at growth rates between 1 and 0.5 monolayers per second at a temperature of 530 °C. To allow the assembling of the QDs the growth was paused for 60 s. The QDs were capped of by 1 nm of GaAs. Afterwards, the chip was annealed at a temperature of 610 °C which causes the QDs to blue-shift [218] due to interdiffusion of the In and Ga. This leads to an emission wavelength of the quantum dots of around 900 nm [219]. The top layer was finalized by an additional growth of 235 nm of GaAs which provides a total thickness of 370 nm for the WG.

7.1.3 Detector Fabrication

The SNSPDs and RF contacts (fig. 7.2a) were fabricated with the process described in section 4.2 of chapter 4. The SNSPDs are patterned to a width of 110 ± 10 nm, a filling factor of 0.55 and a total length of 216 nm. In an additional fabrication step, the NbN film is removed in areas not needed for detectors or RF contacts (fig. 7.2b) to free up space on the substrate for WG integration.

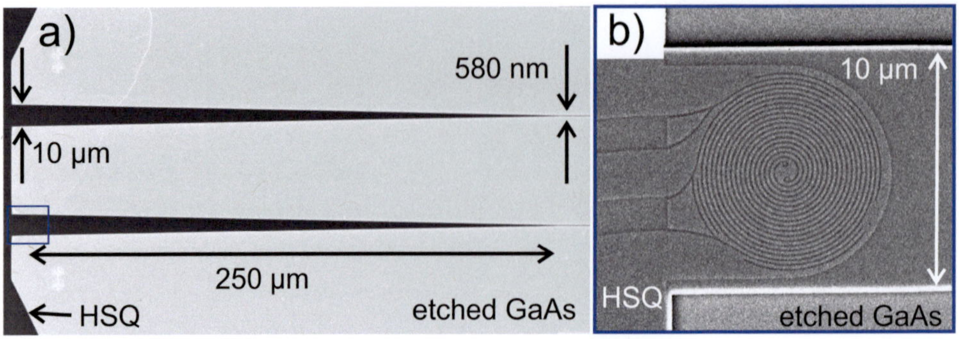

Figure 7.3: (a) SEM image of the WG-based taper (b) WG embedded SNSPD at the end of the taper (coverd with 50 nm HSQ). The dark areas are the HSQ covered areas and the bright areas are the etched GaAs.

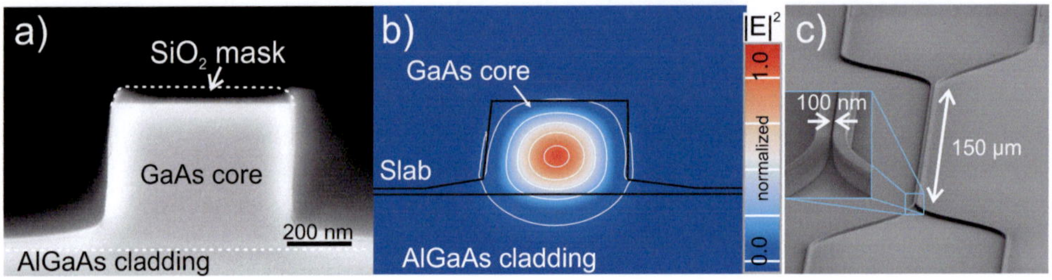

Figure 7.4: a) SEM image of the waveguide cross section after fabrication. b) Simulated TE field for a WG with the used dimensions. c) SEM image of a waveguide-based directional coupler as beam splitter. The inset shows a magnification to the end of the BS and the distance between both WGs in the coupler region. a) and b) are reprinted with permission from U. Rengstl and stem from [110]; c) was reprinted from [67], with the permission of AIP Publishing. All figures were modified.

7.1.4 Waveguide Fabrication

Waveguides and waveguide-based directional couplers were patterned at the Max Planck Institute for Solid State Research in Stuttgart by Thomas Reindl using EBL and etched in a reactive process using an inductively coupled plasma (ICP-RIE) at the University of Stuttgart by Ulrich Rengstl.

To be single mode, the WGs are patterned to a width of 580 nm (fig. 7.4a). In one direction, the WG is tapered in a linear shape over a distance of 250 µm to a final width of 10 µm (fig. 7.3a). In the 10 µm wide section, the full SNSPD is integrated (fig. 7.3b). Hydrogen silsesquioxane (HSQ) was used as a resist, since it offers a high resolution and a good etching stability for the used gases. HSQ is a negative resist that consists of SiH and SiO bonds that are cross-linked to an incompletely closed cage structure. By exposure with an electron beam, SiH bonds, which are weaker than SiO bonds, are broken [220]. The SiO cross-links into a three dimensional network forming SiO_x which is characterized by a high etching durability [220]. The not exposed HSQ is removed by dipping the chip in a 20 % tetramethylammonium hydroxide (TMAH) solution. Afterward, the chip is rinsed with water and ethanol. The HSQ pattern was transferred into a functional WG structure by reactive ion

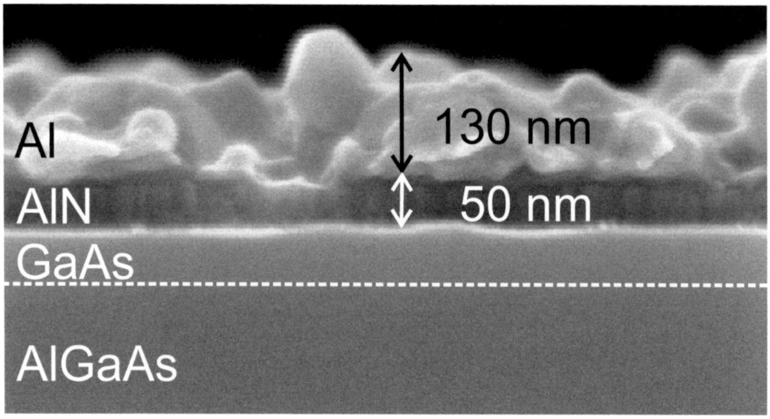

Figure 7.5: SEM image of the crossection of the substrate with AlN/Al cover.

etching with an inductively coupled plasma (ICP-RIE). The etching was done in an ICP-100 system by Oxford instruments at a power of the ICP coil of 350 W using $SiCl_4$ and Ar. The chip was heated to 30 °C to enable an enhanced absorption of the reaction products. The RF etching power was 60 W. The pressure in the recipient was 4 µbar with a gas ratio $SiCl_4$:Ar of 1:1. The WGs were etched as rib waveguides with an etching depth of 320 nm, a 50 nm high slab to a total length of ≈2 mm. In fig. 7.4a, the cross-section of a similar WG is depicted. The GaAs slab prevents the $Al_{0.4}Ga_{0.6}As$ layer from oxidation and the formation of carrier traps due to its surface roughness. Carrier traps in close proximity to the QD can lead to an unstable electrical environment which may lead to a fluctuation of the emission wavelength and intensity of integrated QDs. A simulation of the TE field distribution (fig. 7.4b) done in [110] confirms the single-mode characteristic of a WG in the used material system with the implemented dimensions. By putting two WGs in close proximity to each other, a directional coupler is formed (fig. 7.4c), that can be utilized as a beam splitter (BS) (see section 2.1.4 of chapter 2). For a wavelength of 900 nm, the geometry for a 50/50 beam splitter can be extracted to a coupling length of 150 nm at a gap width of 75 nm. More details on the design and the etching of WGs and beam splitters can be found in [110]. Since the splitting ratio of the BS is very sensitive to fabrication tolerances on the gap width and WG height, each chip containing BSs was fabricated with a variation in the BS gap, to increase the probability of achieving the desired 50/50 ratio.

7.1.5 Al/AlN Cover Layers

To evaluate the origins and the suppression of stray light, an on-chip mirror based on a Al/AlN bilayer was integrated on some samples. The bilayer was sputtered in situ at room temperature using reactive magnetron sputtering. An Al sputter target is used for both films. The AlN layer is hereby sputtered in a N_2^+Ar atmosphere at a thickness of 20 to 50 nm as a transparent dielectric, directly followed by a ≈130 nm thick reflecting Al film in an Ar atmosphere. The resulting bi-layer is seen on top of the etched substrate in fig. 7.5. The AlN film is optimized

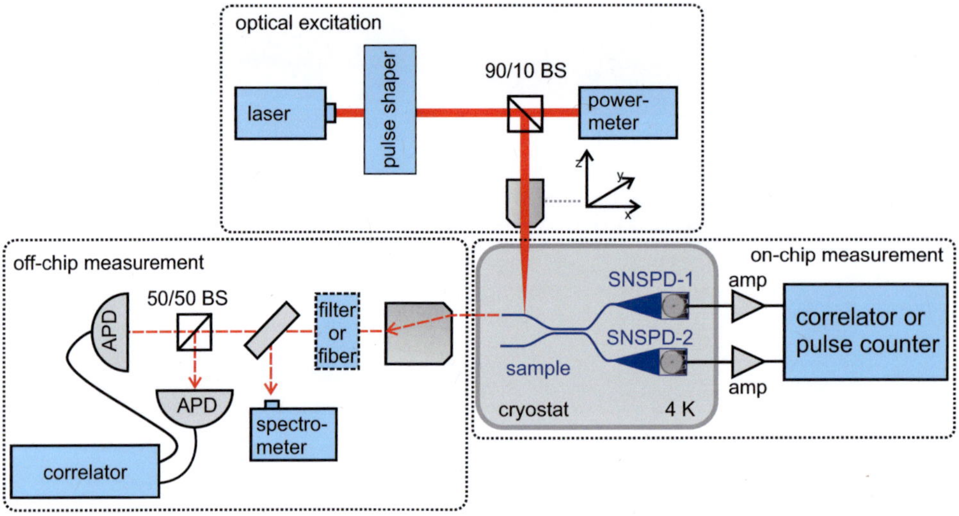

Figure 7.6: Scheme of the optical setup for characterization of a fully on-chip integrated circuit with resonantly-excited quantum dots.

to have a smooth surface (chapter 3) whilst the Al has a granular growth. The deterministic fabrication of Al/AlN bilayers was done using EBL in combination with a liftoff process.

7.2 Experimental Setup

For the thorough characterization of a fully integrated chip, an experimental setup is necessary, which allows the simultaneous excitation of integrated quantum dots, the readout and biasing of SNSPDs, as well as the characterization of optical properties of the integrated WGs and QDs off-chip. A simplified schematic overview of the experimental setup can be seen in fig. 7.6.

Optical Excitation

The sample is optically excited with free space illumination from the top through a microscope objective with a numerical aperture (NA) of 0.45. The objective can be moved in all spatial dimensions using a micrometer stage so any point of the sample can be illuminated deterministically.
Illumination of the sample is done using a titan-sapphire (Ti^{3+}:Al_2O_3) laser pumped by a diode pumped neodymium-doped yttrium orthovanadate (Nd:YVO_4) laser. The laser can be operated under continuous wave (cw), as well as under pulsed excitation with pulses as short as 3.6 ps at a repetition rate of 76 MHz. The emission wavelength of the titan-sapphire laser can be tuned to be between ≈740 nm and ≈930 nm. Emitted photons have a spectral linewidth of ≈100 GHz. The titan-sapphire laser is used in pulsed mode for pulsed-resonant excitation of

the QDs and in cw and pulsed mode for detector characterization. For cw-resonant excitation of the QD, a diode laser is used with an adjustable spectral range of 875 to 940 nm. To enable efficient resonant excitation a pulse shaper is added to the setup. Utilizing the pulse shaper the spectral linewidth of the excitation laser is further reduced and can be adjusted to the resonance wavelength of the QD at high accuracy. The reduced line width of the excitation pulse due to spectral shaping increases its duration significantly: at a linewidth of 11.5 GHz, the pulse duration is 54.6 ps [69]. The light is then sent to a 90/10 BS. 90 % of photons are sent to a power meter to control the optical power and 10 % are sent to the sample. In the focus, the spot size on the sample plane has a diameter of approximately 2.5 µm. During resonant excitation, a low intensity (<1 % intensity of the main pump laser) cw laser with a wavelength of 633 nm is used for gating (see section 2.2.1 in chapter 2) to enable a stable excitation and to gate the emission of the QD [128]. This laser is further regarded as stabilization laser.

Off-chip Measurement

Photons that are guided in the sample and coupled out through the cleaved WG facet are collected from the side by the use of a movable microscope objective with a NA of 0.45. A linear polarization filter and a single-mode fiber can be put inside the beam path to spatially filter out stray light. Photon emission out of a single-mode WG is polarized while stray light is randomly polarized, which is why the polarization filter can reduce the stray light by a factor of two. A fiber in the detection path can be used as a stencil. Correctly aligned only directed photons out of the on-chip WG can couple into the fiber, whilst not directed photons "miss the entry". This allows an effective stray light suppression for off-chip detectors. Afterward, photons can be either sent to a spectrometer or to an off-chip Hanbury-Brown and Twiss setup to investigate the nature of emitted photons. The spectrometer consists of a monochromator that spectrally unfolds the emission. After unfolding the emission is sent either to a CCD chip, which allows an energy resolution of ≈15 µV, or to an avalanche photo diode (APDs). The Hanbury-Brown and Twiss setup consist of a 50/50 BS in combination with two APDs and a time correlating electronics. Further details on the off-chip optical setup can be found in [124].

On-chip Measurement

For on-chip measurements of quantum-dot emission or laser excitation, the integrated SNSPDs are used as detectors. For operation, the SNSPDs are biased using low-noise battery-powered current sources. The detector pulses are separated from the bias current by the use of low temperature, distributed bias tees. The pulses are amplified using a room temperature amplification scheme consisting of three cascaded RF amplifiers: A Minicircuit ZX60-P103LN+ with a typical gain of 15 dB, a noise figure of 0.6 dB and a bandwidth from 40 to 3000 MHz as a first low-noise-amplifier stage and two Minicircuit ZFL-2600VH+ with a typical gain of 20 dB, a noise figure of 5.5 dB and a usable bandwidth from 10 to 2500 MHz as the second and third amplification stage. The total gain is 57 dB with a 3 dB cutoff at 2.8 GHz. The

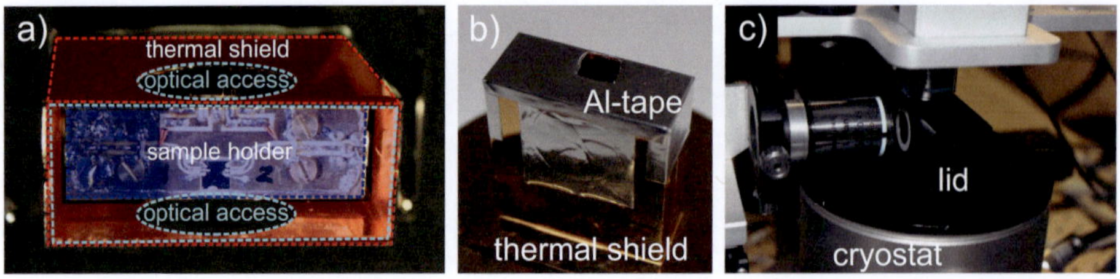

Figure 7.7: (a) Sample holder mounted in the cryostat and thermal shield (colorized red). b) Thermal shield (gold) with Al tape (silver). On the top a square was cut for optical access. c) Closed view of the cryostat. Optical access is possible through either objective through quartz optical windows from the side and from the top.

pulses are either measured with an oscilloscope, a Stanford Research SR620 Time Interval Counter, or a PicoQuant Hydraharp TCSPC electronic.

Cryostat

Since SNSPDs and QDs require sufficient cooling for a successful operation, the basis of the experimental setup is a compact LHe flow cryostat. In the cryostat, the sample is cooled to approximately 4 K. During the experiment the sample is kept in a vacuum at a pressure below 1×10^{-5} mbar. The cryostat is equipped with several DC lines and two RF cables to readout and bias two SNSPDs simultaneously. The sample is mounted on a specially designed sample holder onto the cold finger of the cryostat using Apiezon-N grease (fig. 7.7a) to improve the thermal contact. Piezo-motors inside the cryostat allows the movement of the sample in all three spatial dimensions. During measurements involving SNSPDs, this stage was switched off to minimize noise in the system. Around the cold finger a thermal shield (fig. 7.7a) is placed which is kept at a temperature of 20 K during the experiments. The shield has a rectangular shape and is open from the top. On the long sides optical access is possible through round holes with a diameter of 0.5 inch. To reduce the amount of incoupling thermal radiation, the openings at the top of the thermal shield and on each side of the shield were covered with reflecting aluminum tape (fig. 7.7b). A 5 mm × 5 mm opening was cut into the tape at the top and at one side to provide optical access form the top and from the side. The cryostat is closed by a lid which allows optical access through quartz windows at the top and at the side (fig. 7.7c).

Sample Holder

The sample holder as seen in fig. 7.8 consists of a copper back plate and a Rodgers TMM10i PCB glued onto the back plate using silver conductive epoxy adhesive. The PCB design consists of a coplanar RF line for each detector together with a bias tee with a shared ground connection. The bias tee consists of a 840 nH coil inductor, to block high frequencies in the DC-bias path and a 12 µF SMD capacitor to block the bias current from entering the readout

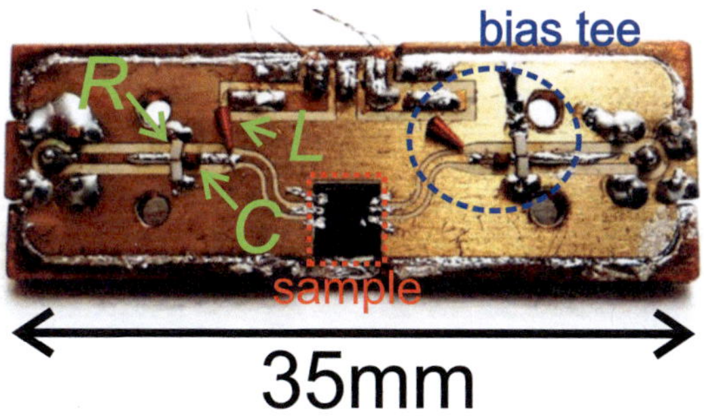

Figure 7.8: The sample holder allows optical access from the top and from the bottom. The position of the sample is marked in red. The right bias tee is marked in blue and the distributed elements (coil inductor L, capacitor C and resistors R) of the left bias tee are marked with green arrows.

path. In addition, two 100 Ω resistors are added to provide a 50 Ω termination for the RF line to omit reflections. The sample is glued using conductive silver paste directly onto the copper back plate to ensure an adequate thermal coupling. Using indium bond wires, the sample is connected to two coplanar RF waveguides. The DC lines for the detector bias are directly soldered onto the PCB. The RF coaxial cables are connected to the PCB with miniSMP connectors at the backside. To ensure a low RF loss the copper layer of the PCB was gold-plated using electrolysis.

7.3 Influence and Reduction of Noise Counts

In previous experiments [108], on-chip operation on a single-photon level could not be demonstrated due to a large amount of introduced stray light. For a successful on-chip operation, a high signal-to-noise ratio is key. False count events are a serious issue and may lead to a non- or malfunctioning of a quantum-photonic device. To reduce the number of noise counts their origins need to be understood. In a fully integrated circuit with laser excited QDs, three main origins of noise counts can be found (fig. 7.9): dark counts (1), thermal radiation (2) and stray light from the excitation laser (3)-(6). (1) All counts that are not caused by photons are regarded as dark counts. They can be distinguished into intrinsic dark counts and counts caused by electronic noise. Intrinsic dark counts can be significantly suppressed by a reducing the operating temperature of the detector [146] and an improved thermal coupling [145]. Counts caused by electronic noise can be avoided by the use of proper shielding of all bias and readout lines, low-pass filters and the use of a low-noise battery-driven current source to bias the detectors. Count events caused by photons are considered as noise counts if they are not stemming from the integrated photon source, in our case the QD. A SNSPD pulse does not give information about the energy and origin of the detected photons. Therefore, all possible sources of photons have to be considered for noise investigations. (2) Thermal radiation is emitted by all bodies as black body radiation with a photon wavelength corresponding to their temperature. In the used free-space setup, photons emitted at a temperature of ≈300 K by black body radiation can enter the cryostat through the quartz windows and are also emitted

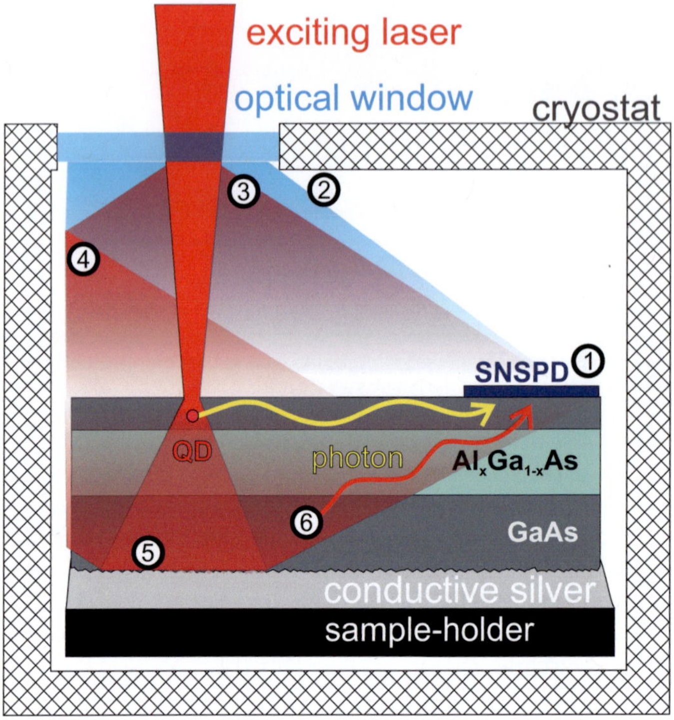

Figure 7.9: Origins of noise counts. 1) Dark counts 2) Thermal radiation 3) Diffraction at the optical window 4) Reflections in the cryostat 5) Reflections at the sample backside 6) secondary emission of photons. The yellow arrow symbolizes the paths for signal photons.

by the quartz window itself. Since a SNSPD can detect photons in the near infrared, thermal emission is able to cause unwanted counts. Thermal photons can be suppressed by blocking all optical paths with a cold thermal shield and by the use of cooled high pass filters in all paths needed for optical access to the SNSPD.

A main source of noise counts for an integrated SNSPD is the laser used for QD excitation, which has an intensity much larger than the intensity of the QD emission. In a QPIC the QD is spatially separated from the detector by a few hundred millimeters to micrometers and all paths on which stray light may reach the detector need to be considered. Laser light can be diffracted and scattered by dust particles and impurities at the optical window (3) or at the cryostat walls (4). After reaching the substrate, the behavior of photons depend on the wavelength. Photons with a wavelength larger than 817.3 nm have an energy smaller than the band gap energy of the GaAs substrate [203] and are able to travel through the substrate at a low loss. They may be reflected at the backside (5) of the substrate or the conductive silver glue and are scattered and may travel to the detector through the substrate. Photons of smaller wavelengths, and consequently with energies higher than substrate band gap are absorbed in the GaAs by exciting electrons into the conduction band of the semiconductor. Since GaAs has a direct band gap, the excited electrons can relax to the valence band by emitting secondary photons (6) which may reach the detector. Direct stray light (3) and photons that reach the

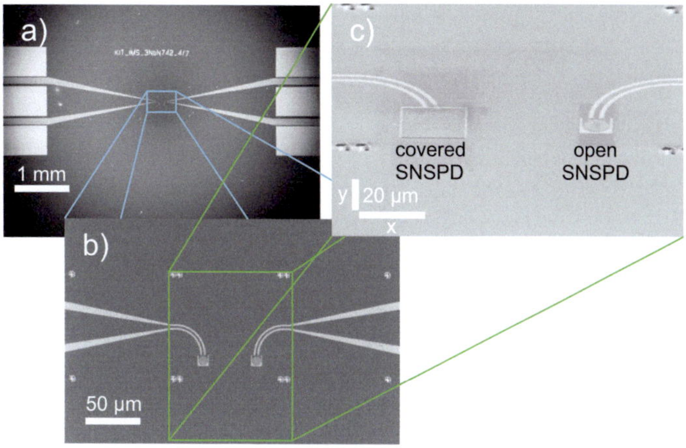

Figure 7.10: SEM images of the sample for investigation of noise counts. a) Overview of the full chip from top. b) Both SNSPDs relative to each other from top. c) Inclined image of one detector with AlN/Al cover on the left and one not covered detector as reference.

detector without travelling through the substrate may be removed by the implementation of a non-transparent cover layer. The probability of photons to be scattered towards the detector can be influenced by the backside topography of the substrate and the excitation angle of the pump laser.
In the following sections, the amount of noise counts is characterized and evaluated regarding their origins. The influence of thermal counts and electrical noise is evaluated by the comparison of the free space to a fiber-coupled setup. The origins and the reduction of stray light is investigated by the deterministic use of Al/AlN cover layers. The influence of the topography of the substrates' backside is examined by the comparison of stray light for single-side and dual-side polished substrates and the laser excitation angle.

7.3.1 Chip Design

To investigate the origins of noise counts, two SNSPDs were placed next to each other at a distance of 50 µm (fig. 7.10). The local proximity of the detectors obviates differences in NbN film thickness and other local variations that may occur during fabrication. The characterization of both detectors is done in the same cooling cycle to provide the best possible comparability. One SNSPD is deterministically covered by a 20 µm × 20 µm Al/AlN patch (fig. 7.10c). The covering was done using 20 nm of AlN as an insulating and transparent spacing followed by a 110 nm thick non-transparent Al-layer.

The cover prevents the illumination from the top but still allows illumination through the substrate. By a variation of the excitation wavelength, the absorption properties of the substrate can be used to modulate the photon flux through the substrate.

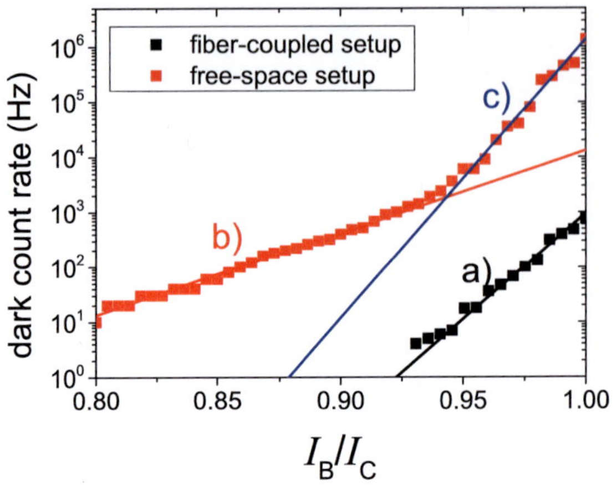

Figure 7.11: Bias-current dependence of the DCR of a SNSPD in the fiber-coupled (black) setup at 4.2 K and the free-space setup (red) at 4 K. The lines are exponential fits to different regions of the bias dependence: a) Exponential fit (black solid line) to the DCR in the fiber-coupled setup. b) Exponential fit (red solid line) to the DCR of the free-space setup below 93 % of the critical current I_C. c) Exponential fit (blue solid line) to the DCR of the free-space setup above 95 % of the critical current I_C.

7.3.2 Setup-Dependent Dark Counts

To investigate the dark and thermal counts, detector dark counts in the free-space setup are compared to dark counts of the fiber-coupled setup (4.3 in chapter 4). In both setups, the detector is operated at a similar temperature: ≈4 K in the free space and 4.2 K in the fiber-coupled setup. The same bias and readout electronic was used. The main difference between both setups, beside from the optical part, is the thermal environment. In the fiber-coupled setup the full sample chamber is at a temperature of 4.2 K. Hence, all bodies with a direct optical path to the detector are at the same temperature as the detector, including the fiber tip. In the free-space setup the detector and sample holder are at a temperature of ≈4 K. Whilst most direct optical paths are blocked by a thermal shield at a temperature of ≈20 K, the two square openings for optical access in the thermal shield with dimensions of 5 mm × 5 mm each face a quartz window at room temperature (≈300 K). In the fiber-coupled setup, dark counts were characterized by blocking the fiber input at the dipstick. In the free-space setup dark counts were measured using conditions comparable to the conditions of a fully integrated experiment. The measurement takes place in a dark room with excitation lasers blocked. A cloth is placed around the cryostat, to block all but the optical paths that are used for QD excitation and observation.

The dark count rates were measured in dependence of the bias current in both setups using a single, non covered SNSPD (fig. 7.11). In the fiber-coupled setup a mono exponential increase of the DCR with increasing bias current can be seen (a), which fits very well with the behavior expected for intrinsic dark counts [146]. In the free-space setup, the DCR is more than 2 orders of magnitude higher. Two different dark count regions are visible: at bias currents below 93 % the DCR follows an exponential increase (b) with the bias current.

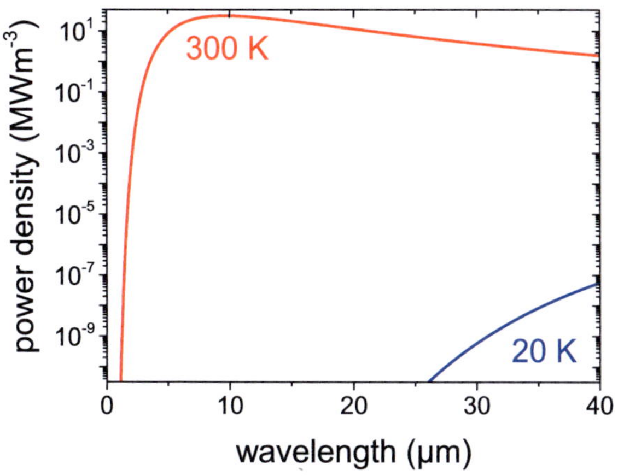

Figure 7.12: Calculated spectral power density at 300 K (red) and 20 K (blue) using Planck's law.

Close to I_C a stronger exponential dependence is observed (c). The two different inclinations indicate two different causes for the observed dark counts. The dependence of the dark count rate close to the detectors I_C is associated with the intrinsic dark counts of the detector. The intrinsic dark count rate of the SNSPD in the free space cryostat (c) is significantly higher than what was seen in the fiber-coupled cryostat (a). Since for both cases the SNSPD is at a similar operation temperature, an increase in the intrinsic DCR can be caused by a weaker thermal coupling, by vibrations or by temperature fluctuations in the flow cryostat. Since the main differences in both setups is the thermal environment of the detector, the dark counts with dependence (b) are probably caused by thermal emission. Direct sources for thermal radiation in the free-space setup have to be considered: the 20 K thermal shield or thermal emission at 300 K by the quartz windows and thermal emission that may enter through the quartz windows.

The spectral power density of a black body radiator can be calculated using Plank's law and is shown in fig. 7.12. For a black body with 300 K the emission of radiation starts in the near infrared. The thermal emission at 20 K starts in the mid-infrared with a power density orders of magnitudes smaller in comparison to the emission at 300 K. Whilst the SNSPD is in the probabilistic detection regime, which explains the exponential dependence, it will still detect photons in the near and mid infrared. However, since the thermal emission at 20 K has a considerably lower power density and occurs at longer wavelengths, where the detector has a lower DE, it can be concluded that the thermal shield does not significantly contribute to the measured dark counts.

In summary, in the free-space setup the DCR is significantly increased due to thermal radiation stemming from the room-temperature optical windows and laboratory environment. Close to I_C an increase of the intrinsic DCR may be caused by weaker thermal coupling, vibrations or temperature fluctuations in the flow-cryostat. An optical high pass filter at the 20 K thermal shield should be able to significantly reduce the amounts of thermal dark counts.

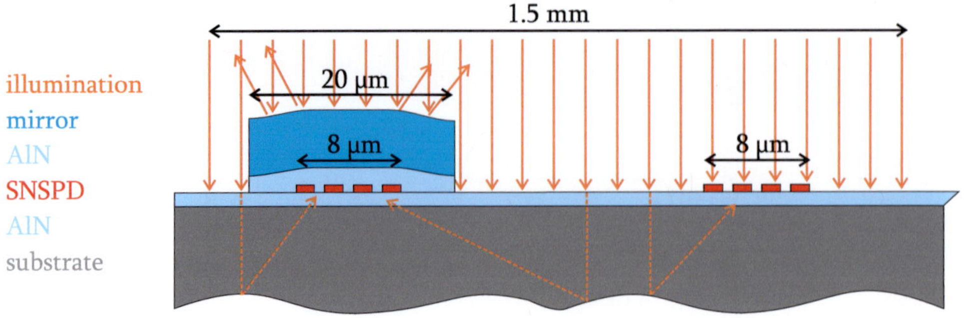

Figure 7.13: Sketch of the illumination of the detector area using a broad beam.

7.3.3 Stray Light

Since stray light from the excitation laser is one of the main sources for noise in a QPIC, the influence of stray light is quantified and the origins are evaluated. The investigation is done using top illumination of the sample at different positions relative to the detectors in the following section. External photon sources with two different spot profiles are used: a broad beam with a spot diameter of ≈1.5 mm and a focused spot with a diameter of ≈2 µm. The broad spot illumination simulates the influence of spatial distributed stray light on-chip and allows the evaluation of the effectiveness of a detector cover to reduce stray light from spatially distributed photons. A focused spot allows the simulation of stray light of an excitation laser during QD operation. The stray light is investigated in dependence of the excitation distance to the detector by a spatial movement of the excitation spot on the sample. By the use of single-side polished as well as dual-side polished substrate, the influence of the backside topography of the substrate on stray light is investigated. The unpolished backside with a peak to peak roughness of ≈10 µm of the single-side polished sample offers a variety of reflection angles and a diffuse reflection is expected. The dual-side polished substrate has an atomically flat polished back surface, which may allow a more directed reflection. A tilt of the excitation laser is investigated as a method to reduce stray light on the SNSPD by directed scattering of the excitation laser away from the detector.

Broad-Beam Illumination

The investigation with broad beam illumination was performed using the fiber-coupled setup described in detail in section 4.3 of chapter 4. In the beginning of each experiment, the fiber is centered on top of both detectors perpendicular to the substrate. A sketch of the sample illumination is shown in fig. 7.13. The use of a multimode fiber with a core diameter of 105 µm and a numerical aperture NA of 0.22 at a distance of more than 3.5 mm to the substrate, leads to a broad illumination spot with a diameter of at least 1.5 mm on the substrate. This diameter is much larger than the diameter of the detector (8 µm) and the distance between the detectors (50 µm), so both detectors are homogeneously illuminated. The sample is characterized over a spectral range from 400 to 1700 nm, with a focus on 900 nm as the main wavelength of the

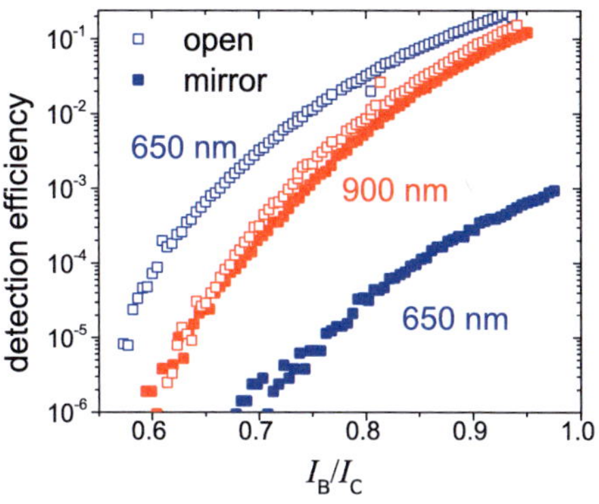

(a) single-side polished

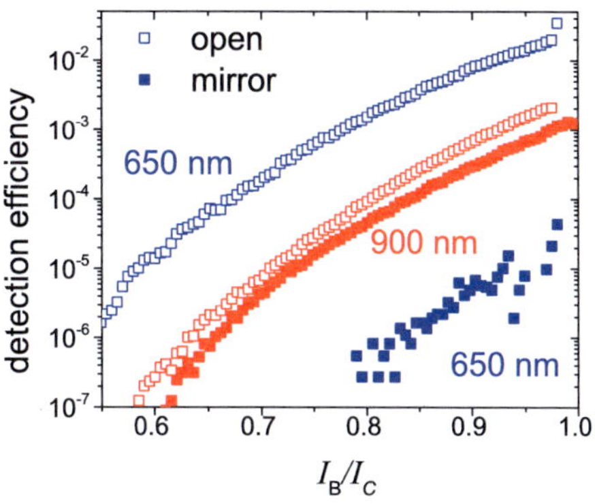

(b) dual-side polished

Figure 7.14: Detection efficiency in dependence of the bias current for the SNSPD with (closed-symbols) and without (open-symbols) cover mirror. Blue shows the data for $\lambda = 650\,\text{nm}$ and red for $\lambda = 900\,\text{nm}$.

QDs. The absorption characteristics of GaAs can be used to evaluate the amount of photons that reach the detector through the substrate and from the top and allows the evaluation of the blocking efficiency of the top cover.

The bias dependence of the DE is shown in fig. 7.14 for a (a) single-side polished and a (b) dual-side polished substrate. The detector with cover mirror is depicted in open symbols and the detector without in closed symbols. The bias dependence was recorded at 650 and 900 nm. At a wavelength of 650 nm the photon energy exceeds the band gap energy and the photons are absorbed in the substrate. At a wavelength of 900 nm the substrate is transparent. The DE increases with increasing bias current. The DE is higher for the open SNSPDs. At a wavelength of 900 nm the efficiency has the same order of magnitude for the open, as well as the covered detector. At a wavelength of 650 nm the covered detector has a suppressed DE in comparison to the open detector. DE for the closed SNSPD is reduced by a factor of 700 on the single-side polished substrate (fig. 7.14a) and by a factor of 1700 on the dual-side polished substrate (fig. 7.14b).

The wavelength dependence of the DE was measured for $I_B = 0.9 I_C$ and is shown in fig. 7.15. For open detectors, the DE decreases with increasing wavelengths. At wavelengths longer than the direct bandgap of GaAs, the substrate gets transparent and the detection efficiency increases by 70 % for the single-side polished substrate and 32 % for the dual-side polished substrate. The DE of the covered SNSPDs is several orders of magnitudes lower than the DE of open SNSPDs as long as the photon energy is smaller than the bandgap energy. For higher energies, at wavelengths longer than 818.3 nm, the DE increases to a level in the same order of magnitude as for the open detector. The suppression of counts for photon energy above the bandgap in the order of three to four magnitudes confirms the high reflectivity of the cover mirror. The DE dependency in the range of 400 to 500 nm for the covered detector could be caused by the radiative recombination of excitons generated by photo-excitation with a higher probability towards higher photon energies.

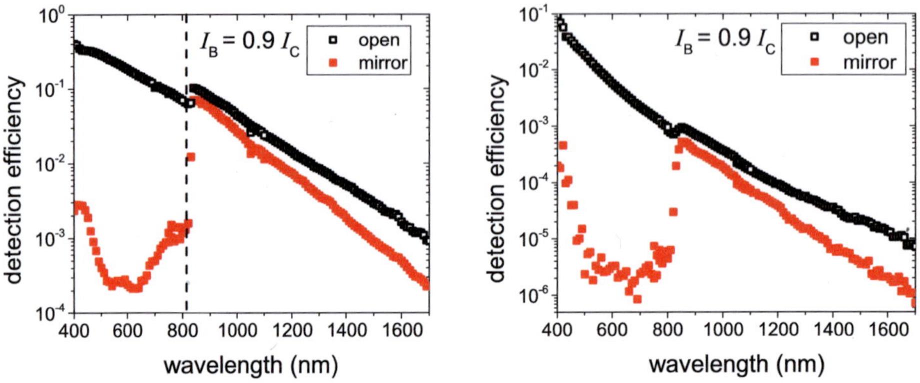

(a) single-side polished

(b) dual-side polished

Figure 7.15: Spectral dependence of the DE with (closed red symbols) and without (open black symbols) Al-cover mirror. For SNSPDs on single-side polished substrate in (a) and on dual-side polished substrate in (b). At the band gap energy of GaAs a strong change in detection efficiency is visible.

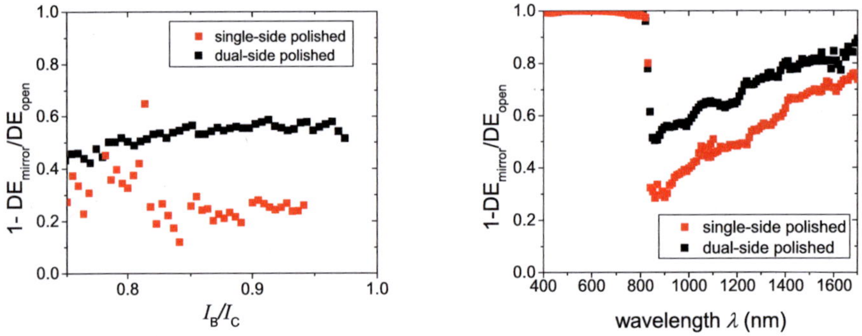

(a) bias dependence at $\lambda = 900$ nm

(b) wavelength dependence

Figure 7.16: Suppression of counts for the single-side polished (black) in comparison to the dual-side polished substrate (red) by the cover mirror for the bias dependence (a) at 900 nm and the spectral dependence (b).

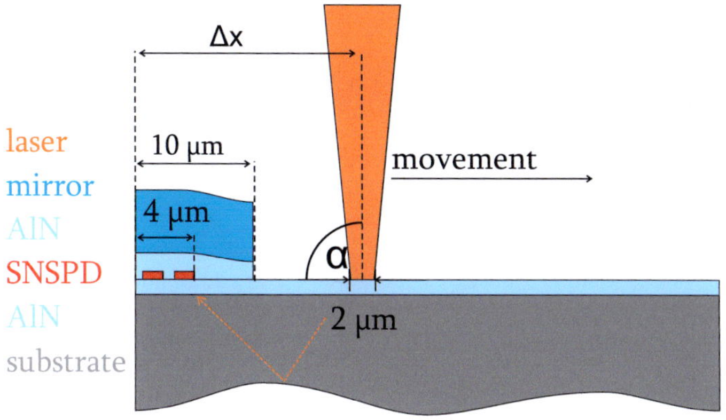

Figure 7.17: Measurement with focused beam illumination.

To allow a direct comparison, the suppression of the DE by the cover mirror was calculated as $1 - DE_{mirror}/DE_{open}$ (fig. 7.16). DE_{mirror} is the DE of the covered SNSPD and DE_{open} of the open SNSPD. For both substrates the DE suppression by the cover in dependence of the bias current is compared in fig. 7.16a at $\lambda = 900\,nm$. The spectral dependence of the DE suppression by the cover for both substrates is compared in fig. 7.16b. On the single-side polished substrate the count rate is suppressed by ≈20 % and on the dual-side polished substrate by ≈50 %. In the spectral dependence, short wavelength stray light is efficiently suppressed on both substrates. This proves a high efficiency of the applied cover mirror to block the direct light path to the detector. For wavelengths slightly longer than the bandgap wavelength the DE suppression is ≈40 % stronger for the dual-side polished in comparison to single-side polished substrate, which decreases to ≈15 % at 1700 nm. A suppression of only 20 % (50 % for DP) by the mirror for photon energies below the bandgap suggests that the predominant number of photons reach the detector indirectly through the substrate. This shows the importance of blocking photons from entering the substrate. The improved suppression by the use of a dual-side polished substrate suggests that the reflexion properties of the substrate backside interface play an important role in reducing spatially distributed stray light.

Focused-Beam Illumination

The response of the open and covered SNSPDs was investigated with focused beam illumination at 900 nm. During the focused beam illumination experiment, the samples were operated in the free-space setup (section 7.2). The focused laser spot has a round Gaussian spatial distribution on the sample with a diameter of 2 µm at which point the intensity drops to $1/e^2$ of its intensity at the peak. The illumination is done perpendicular to the substrate surface at different distances to the SNSPD (fig. 7.17). The laser was focused on the sample surface, centered on the corresponding detector for reference and than moved away. The benefit of the detector cover and substrate choice is evaluated. In the end of this section, the effect of

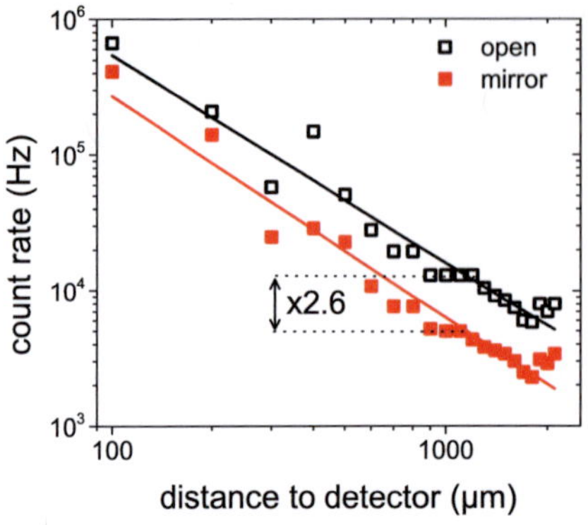

(a) Cover influence

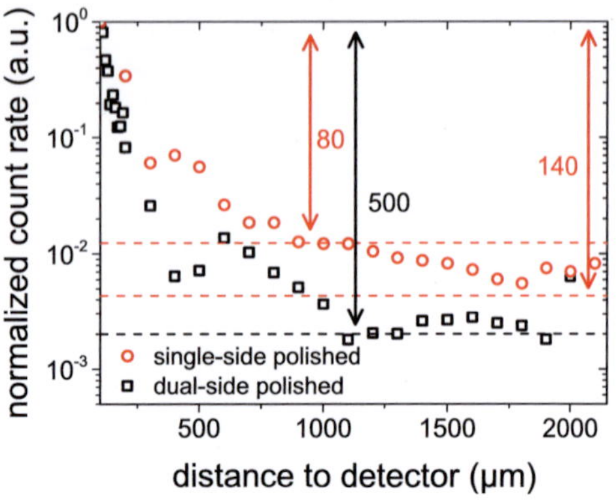

(b) Substrate incluence

Figure 7.18: Stray light on a SNSPD, in dependence of the excitation distance. (a) Count rate comparison for an open (black open symbols) and a covered detector (red closed symbols). Solid lines are exponential fits. (b) Count rate comparison for an open SNSPD on the single-side polished (red circles) and dual-side polished (black squares) substrate. The dashed lines are to guide the eye.

a modified excitation angle α of the pump laser is investigated for SNSPD on a dual-side polished substrate.

The stray light count dependence on the excitation distance to the detector from 100 to 2000 µm is shown with and without a cover mirror in fig. 7.18a. The investigation was done at $I_B = 0.9 I_C$. The data shows an exponential dependence up to a distance of ≈1600 µm. For longer distances the count rate stays at an almost constant level with some fluctuations in dependence of the exact position of the beam. Stray light is suppressed by a factor of 2.6 at a distance of 1000 µm by the use of the Al/AlN cover. From an exponential fit to the distance-dependent count rate, an attenuation of stray light in dependence of the excitation distance can be extracted to 23.8 dB/mm for the open detector and 25.4 dB/mm for the closed detector. The count dependence on distance for the single-side polished substrate in comparison to the dual-side polished substrate is shown for SNSPDs without cover in fig. 7.18b. The count rates were normalized to unity at a distance of 100 µm. The count rates decrease exponentially until they level out. For the dual-side polished substrate the count rate levels out at a distance of 1000 µm, while a further decrease is seen for the single-side polished substrate. At a distance of 1000 µm the count rate is reduced by a factor of 80 for the single-side polished substrate compared to a factor of 500 for the dual-side polished substrate. The exponential decay for small distances in combination with a flat count rate dependence on distance for distances larger than 1000 µm indicates two mechanisms for stray light to reach the detector. The first is dependent on the distance of the excitation laser to the detector, the second is spatially independent. Scattered light inside the chamber is a strong candidate for the second mechanism. Since the movement of the laser is small in comparison to the size of the sample chamber, it should be independent of the excitation position on the substrate.

To further reduce the amount of stray light seen by the detector, a slight incline of the excitation laser was tested on the open SNSPD on the dual-side polished substrate. The idea is that

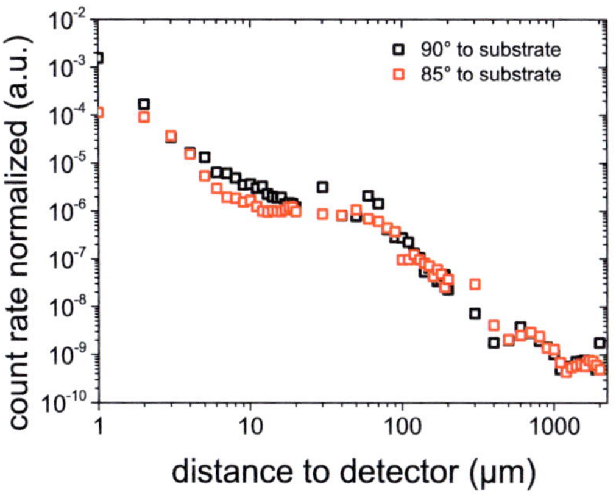

Figure 7.19: Differences in the stray light count rate for laser illumination at an angle α of 90° and an angle of 85° to the dual-side polished substrate. The count rate was normalized on the laser power.

the polished backside may be used as a "mirror" to reflect photons from the excitation laser away from the detector. Since the angle of incidence is equal to the angle of reflection, an inclination of the excitation laser relative to the backside plane is necessary to scatter stray light in a defined direction. As a test, the laser inclination is changed by 5° (α = 85° fig. 7.17) from its previously perpendicular position in relation to the backside plane and compared to the original position (fig. 7.19).
Close to the detector the stray light is reduced by the inclination. However, at a distance larger than 60 µm from the detector no influence of the inclination on stray light can be seen.

In conclusion, to minimize noise counts on the detector, several parameters of the experimental setup can be optimized. Ideally, a fiber-coupled setup in combination with shielding for thermal radiation should be used to minimize dark counts. Stray light affecting the detector can be efficiently reduced by the use of a direct cover on top of the detector and by utilizing a dual-side polished substrate. Other modifications might prove even more efficient, like an anti-reflection coating on the backside in combination with an absorber, allowing for photons to exit the substrate at the backside of the sample and be absorbed. A small inclination of the excitation laser away from the detector reduces stray light if the QD is close to the detector, but is not efficient at longer excitation distances. However, this does not conclude that a larger angle is not efficient in reducing stray light. However, larger angles may not be feasible for an efficient excitation of QDs. Furthermore, two different mechanisms for stray light were found. One is position-dependent and can be significantly reduced by exciting QDs at a distance larger than 1 mm to the detector, and a second, weaker one is a position-independent mechanism.

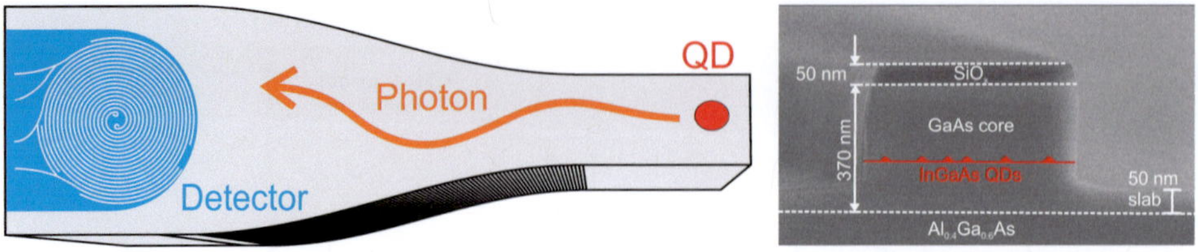

(a) Detector on WG with taper

(b) SEM of the WG faced

Figure 7.20: a) Sketch of a WG-SNSPD with WG taper. b) SEM of the WGs faced shows the WG cross-section. The position of the self-assembled QDs inside the WG is indicated by the red line.

7.4 Waveguide Integrated SNSPDs and QDs

To make the next step towards a fully integrated circuits, SNSPDs were integrated onto straight waveguides along with QDs. A sketch of the utilized design is shown in fig. 7.20a. A 580 nm wide single-mode waveguide is patterned onto a sample with embedded QDs. The SNSPDs are integrated onto the WGs after a 250 µm long taper. The samples were mounted inside the free-space setup for a full characterization. Photons can be excited in the WG, guided and detected by the on-chip SNSPD or coupled out at the other side of the sample through a cleaved faced and characterized off-chip. A SEM of the cleaved faced is shown in fig. 7.20b. This enables the investigation of the visible brightness of the integrated QDs, the absorption properties of the used WG and the amount of laser stray light coupling into the WG on- and off-chip. The on-chip generation and detection of photons is demonstrated. Using the QD brightness in relation to counts caused by noise sources, a SNR can be extracted. The SNR is investigated in respect to the excitation distance to the detector and characterized for resonant and above-band excitation of the QD. The laser excitation is performed perpendicular to the substrate surface from the top and is, therefore, also perpendicular to the WG. By exciting perpendicular to the WG, most laser light should be reflected perpendicular to the waveguide and should not propagate inside the WG [68].

7.4.1 Above-Band Excitation and WG Absorption Characteristics

The excitation of WG-integrated QD was investigated off-chip using the spectrometer. The QDs were excited using above-band excitation with a focused spot at $\lambda = 800$ nm. With an excitation wavelength at an energy significantly larger than the QDs energy level, QDs in the whole laser spot are excited (see: section 2.2.1). The randomly excited QDs together form a QD ensemble. Emitted photons are guided inside the waveguide and couple into free space. They are collected by a microscope objective and guided to a spectrometer. The measured spectral intensity of the QD photo luminescence (PL) is shown in fig. 7.21a. By moving the spot on the WG, QD ensembles are excited at different distances to the cleaved edge. The emission of excited QDs is in the range of 865 to 895 nm, with a main number of QDs emitting at a wavelength of 872.5 ± 7.5 nm. The PL intensity over the investigated spectrum is summed up and plotted in dependence of the distance to the WG edge in fig. 7.21b. The scatter of

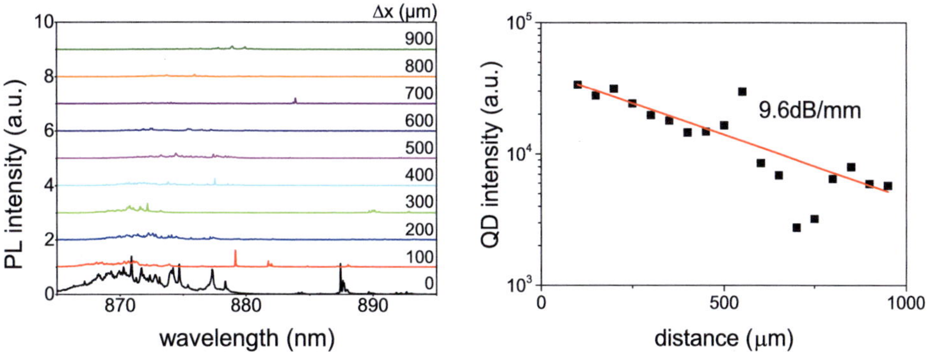

(a) QD PL intensity spectrum

(b) QD intensity on distance

Figure 7.21: Above-band excitation of QDs in a WG at different distances to the cleaved edge. (a) shows the PL intensity spectrum for different distances to the cleaved edge. The distance is indicated in μm on the right side of the graph. (b) accumulated intensity in dependence on the traveled distance of photons in the WG. The red line is an exponential fit to extract the attenuation.

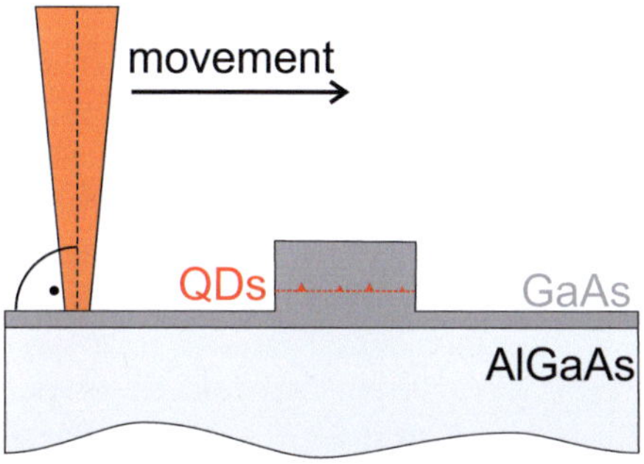

Figure 7.22: Cross section of the WG to visualize of the move path for the excitation laser.

measured values is caused by a spatially dependent number of excited QD and a changing excitation efficiency for the corresponding QD ensemble at each position. The QD density is not identical at each position and their polarization does not always fit to the polarization of the excitation laser. Out of an exponential fit to the data, the optical attenuation of the WG is extracted to be 9.6 dB/mm. This value is rather high compared to state of the art absorption values of 2.6 ± 0.6 dB/mm [67] for GaAs WGs in the investigated spectral range. A higher surface roughness of the WG can be a possible reason for the higher attenuation.

7.4.2 On-Chip Emission and Detection of Photons

The detection of QD photons was investigated on-chip using integrated waveguide SNSPDs (WG-SNSPD). WG-SNSPDs together with WG-integrated QDs allow the characterization of the on-chip signal-to-noise-ratio by deterministically "toggling" the QD emission. The coupling and guiding of laser light inside the WG is characterized by moving the excitation

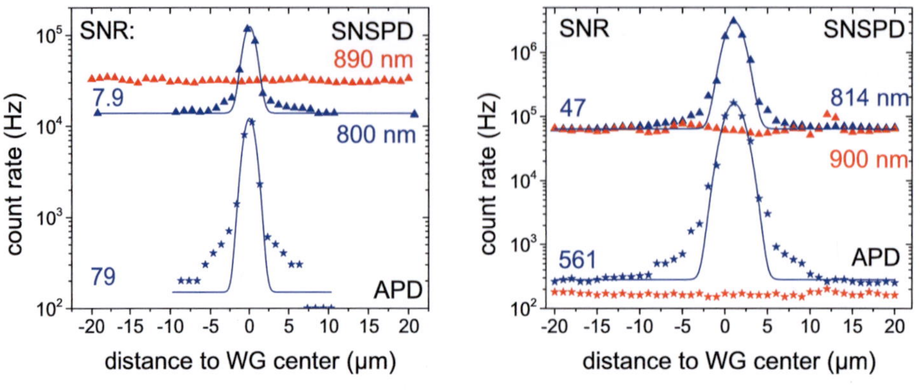

(a) single-side polished substrate

(b) dual-side polished substrate

Figure 7.23: Count rate in dependence on the position of the excitation laser. The dependence for above-band excitation (800 nm & 814 nm) is shown in blue, and for below-band excitation (890 nm & 900 nm) in red. The count rate for the SNSPD is shown in triangles and the count rate for the APD in stars. Solid lines are fits to the measured data.

laser across the WG. The experiment was performed for a wavelength above-band at $\lambda \approx$ 800 nm to excite QDs and below-band at $\lambda \approx$ 900 nm to investigate stray light without QD excitation. The SNR for an open SNSPD on a single-side polished substrate is compared to the SNR of a SNSPD utilizing a cover mirror on a dual-side polished substrate.
The experiment is sketched in fig. 7.22. First, the laser was focused on the substrate at a distance of 20 µm from the WG and then moved perpendicular across the WG. The distance to the SNSPD during measurement was 1200 µm for the single-side polished substrate (fig. 7.23a) and 300 µm for the dual-side polished substrate (fig. 7.23b). The distance to the cleaved WG facet was 50 µm in both cases. The photons are guided in the waveguide towards the integrated SNSPD and in the opposite direction towards the cleaved WG faced where they are coupled into free space to an APD. In front of the APD a long pass filter with a cutoff at 865 nm was placed to filter out the laser light during above-band excitation.

When moving the laser at an above-band wavelength towards the WG, the count rate is flat until it starts rising at a distance of ≈5 µm from the WG center and reaches a maximum in the middle of the waveguide. The behavior repeats in reverse symmetry when moving off the WG. The flat count rate at a distance is caused by stray light counts. When the light spot of the excitation laser overlaps with the WG, QDs are excited in the waveguide causing a strong rise of the count rate. For below-band excitation, the count rate stays flat: no additional stray light can be seen when the exciting laser is on top of the WG. Hence, no coupling of laser photons into the WG is observed. To calculate the signal-to-noise ratio, the maximum count rate was extracted from a fit to the measured data points. For the fit we assume that the QDs are homogeneously distributed in the WG, their coupling efficiency to the waveguide is position-independent and they will only be excited by direct laser light. The count rate of the detector is directly proportional to the number of excited QDs with an offset by stray light. The emission of QDs is directly proportional to the area of the laser spot that overlaps with the WG, weighted by its intensity profile. Since the diameter of the laser spot d_{laser} is way smaller than the length l_{WG} of the WG, the WG can hereby be resembled by a one dimensional rectangular function $\sqcap(x)$, and the position dependence of the count rate can be expressed as:

$$f_{WG}(x) = \eta_{count} \cdot \sqcap(x) = \begin{cases} 0 & \text{when } |x| > \frac{w}{2} \\ \eta_{count} & \text{when } |x| \leq \frac{w}{2}, \end{cases} \tag{7.1}$$

where x is the position of the laser spot relative to the middle of the WG and w is the width of the WG. η_{count} is the probability of a laser photon exciting QD photon emission that is detected by the SNSPD. The laser spot is assumed to have a 2 dimensional Gaussian shaped intensity distribution on the substrate. As $d_{laser} << l_{WG}$, the laser spot can be expressed as a one-dimensional Gaussian distribution:

$$f_{laser}(x) = N_{photons} \cdot \frac{1}{\sqrt{2\pi\sigma_{laser}^2}} e^{-\frac{x^2}{2\sigma_{laser}^2}}, \tag{7.2}$$

where $N_{photons}$ is the total number of photons in the laser spot, and σ_{laser} is the standard deviation. The fit function is given by the convolution of the rectangular function with the Gaussian distribution with a noise offset y_0:

$$f_{fit} = N_{noise\ counts} + (f_{WG} * f_{laser})(x) = N_{noise\ counts} + \int f_{WG}(\iota) f_{laser}(x - \iota) d\iota \tag{7.3}$$

$$f_{fit} = N_{noise\ counts} + \int_{-\frac{w}{2}}^{\frac{w}{2}} \eta_{count} \cdot N_{photons} \cdot \frac{1}{\sqrt{2\pi\sigma_{laser}^2}} e^{-\frac{(x-\iota)^2}{2\sigma_{laser}^2}} d\iota. \tag{7.4}$$

For fitting, the count rate at a distance of 20 µm from the WG core was used for $N_{noise\ counts}$ and the width w was set to the width of the WG, which is 580 nm. The fit function fits very well to the measured peak, however, it fails to include the observed initial rise of the count rate close to the WG. This count rate increase can be explained by the presence of an airy disk around the laser spot caused by diffraction. Since the intensity of the airy disk is orders of magnitude lower, the used fit function nevertheless allows an accurate extraction of the peak height and width of the spot size of the excitation laser. The diameter of the laser spot was extracted as the diameter where the area of the Gaussian profile outside the diameter reaches $1/e^2$ of its overall area. For the measurement on the single-side polished substrate, the spot diameter is extracted to 2.1 ± 0.1 µm and on the dual-side polished substrate to 2.7 ± 0.2 µm. The SNR was calculated as the number of detection events caused by photons emitted by the QDs ($N_{QD\text{-}photons}$) in relation to the measured $N_{noise\ counts}$. $N_{QD\text{-}photons}$ is determined by subtracting $N_{noise\ counts}$ from the total number of detection events ($N_{total\ counts}$).

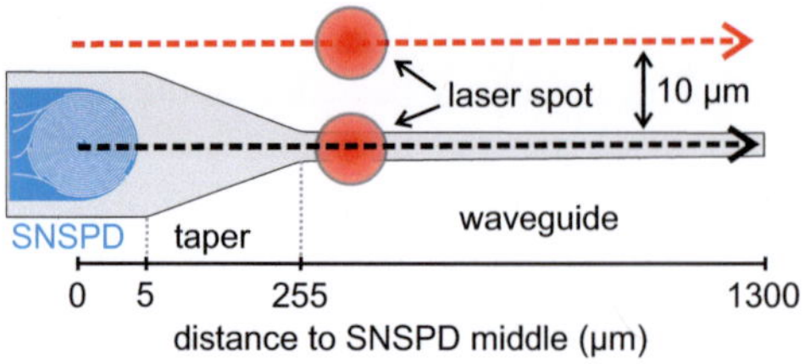

Figure 7.24: Above-band laser excitation experiment for different positions of the laser spot on and off the waveguide. The dashed line shows the move path of the laser relative to the WG.

$$SNR = \frac{N_{\text{QD-photons}}}{N_{\text{noise counts}}} = \frac{N_{\text{total counts}} - N_{\text{noise counts}}}{N_{\text{noise counts}}} \tag{7.5}$$

The SNR for the WG-SNSPD on the single-side polished substrate is 7.9 and 79 for the off-chip APD. With the dual-side polished substrate a SNR of 47 is reached for the SNSPD and a SNR of 561 is reached off-chip at the APD. After subtraction of dark counts of the SNSPDs used, the on-chip SNR is calculated to 8 for the single-side polished substrate and to 58 for the dual-side polished substrate. The higher SNR of the APD is associated with a lack of stray light. However, the absolute count rates and the corresponding detection efficiencies are significantly higher for the WG-SNSPDs. The SNR of the SNSPDs cannot be directly compared since the waveguide is shorter in the dual-side polished substrate case. A larger SNR for the detector with Al-cover on a dual-side polished substrate is in good agreement with the findings in section 7.3.3. The considerable difference in the SNR measured off-chip between both samples is caused by a higher excitation power of the laser for the measurement on the dual-side polished sample. In addition, the spot diameter is larger, which means more QDs are excited. The measured SNR would lead to a theoretical $g_2(0)$ value for single-photon emission of 0.212 (0.209 without DCR) for the single-side polished substrate and of 0.04 (0.03 without DCR) for the dual-side polished substrate. This is sufficient for an on-chip experiment on a single-photon level. However, an ensemble of QDs was excited in this case and not a single-QD. The emission intensity for a single QD is expected to be lower, which decreases the SNR. The lack of laser photons coupled into the WG shows that a WG-integrated spectral filter would not improve the SNR.

7.4.3 Optimal Excitation Distance

For a SNR as high as possible, the optimal excitation distance of the QDs to the SNSPDs needs to be found. While stray light was shown to decrease (section 7.3.3) with a larger distance to the detector, until it reaches a spatially independent base level, QD emission will be attenuated inside the WG in dependence on the WG length. Consequently, an optimal excitation distance is expected. To evaluate the optimal excitation distance, QDs were exited above-band inside the WG at different distances to the SNSPD. The experiment was performed using a WG-SNSPD on a single-side polished substrate without Al cover. The experiment

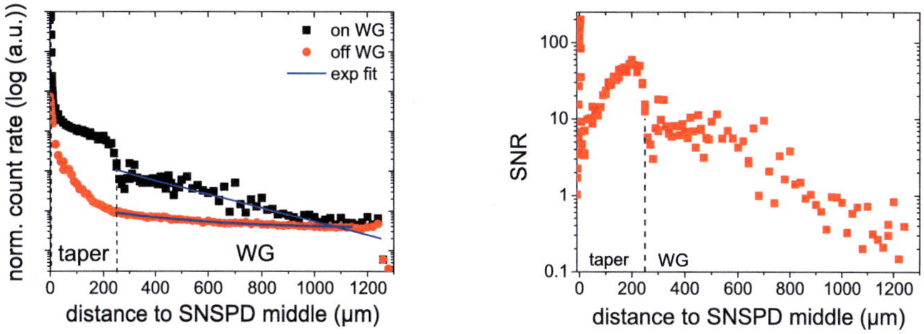

(a) Excitation of QD ensemble and stray light along the WG

(b) SNR along the WG

Figure 7.25: Count rate and SNR dependence of the excitation distance to the WG-SNSPD. (a) The count rates are normalized to the excitation power and depicted without dark counts. The black squares shows the count rate dependence while exciting QDs in the WG, the red circles the count rate while moving the laser in parallel to the WG at a distance of 10 µm. The blue solid lines are exponential fits to the data. The dashed lines separates different parts of the WG structure (see fig. 7.24). (b) shows the SNR ratio calculated from (a).

is sketched in fig. 7.24. For the investigation, the laser was set to a wavelength of 800 nm and positioned in the middle of the SNSPD. It was then moved away step by step from the SNSPD on top of the taper and the WG. As long as the laser spot sits on the taper or the WG, QDs are excited and the Qd-emission is observed by the on-chip SNSPDs. To characterize the dependence on distance for stray light, the same experiment was repeated in parallel to the taper and WG, at a distance of 10 µm, by moving the laser spot over the substrate.

During measurement the laser power was adjusted to keep the count rate in the single-photon regime on the SNSPD. The count rate normalized on the excitation power, in dependence on the laser position is shown in fig. 7.25a. To be able to compare the measured data points for different excitation powers, the DCR was subtracted. Whilst moving the excitation laser on the WG, the count rate behaves differently in dependence on the underlying WG part. In the first part, the SNSPD is directly illuminated by the laser spot and the detector counts are mainly caused by direct laser light. The count rate at this point is proportional to the area overlap of the SNSPD with the laser spot and a flat dependence is visible while the full spot sits on the detector. The second part is the linear WG taper. The taper starts at a distance of 5 µm from the middle of the detector and ends at a distance of 255 µm. At the start the taper has a width of 10 µm and photons in the full area of the 2.1 ± 0.1 µm wide spot excite QDs in the taper. At the end of the taper it has a remaining width of 580 nm, so only the inner part of the spot excites QDs. Furthermore, the local loss of photons in the taper changes because of the changing width: for a broader width, absorption effects at the etched side walls are less relevant. The third part is the 580 nm wide WG. The change in count rate in the WG region reflects the attenuation in the WG and follows an exponential dependence with an attenuation of 15.48 dB/mm. However, a large scatter in the observed count rate is present, caused by the random distribution of QDs in the WG. The QD density fluctuates over the chip and the coupling efficiency to the WG is dependent on the QD position and its polarization. A drift of the laser perpendicular to the WG during the measurement could further influence

the scattering. Because of the small width of the WG a small drift of the laser can have a significant effect on the effective area of the excitation spot on the WG. At a distance of 1260 µm from the middle of the SNSPD, the count rate sharply drops since the excitation laser drives of the end of the sample.
The stray light dependence recorded with excitation in parallel to the WG shows an exponential dependence on the excitation distance with an attenuation of 3.5 ± 0.2 dB/mm until a distance of $\approx$1100 µm to the middle of the detector where the count rate flattens out. The count rate for stray light sharply drops when the laser spot moves off the sample at distance of 1260 µm to the middle of the SNSPD. The SNR calculated in dependence on the distance is shown in fig. 7.25b. The SNR increases along the taper until the width of the taper gets smaller than the width of the laser spot. The SNR then drops significantly until the taper reaches the width of the WG. When on the WG, the SNR decreases with increasing distance to the SNSPD. For the characterized sample, the length of the WG should be kept as short as possible to ensure a high SNR. The reason for this is the high attenuation of the photon intensity in dependence on the length of the WG with 15.48 dB/mm in relation to the attenuation of the stray light in dependence on the distance to the SNSPD with 3.5 ± 0.2 dB/mm. Absolute SNR values for further measurements cannot be concluded as the WG absorption measured is significantly higher than for state of the art WGs (2.6 ± 0.6 dB/mm [67]) and, in addition, the SNR was measured for a QD ensemble and not for a single QD. A WG with an attenuation below 3.5 ± 0.2 dB/mm would allow for an improvement in SNR for a longer excitation distance. In this case the best SNR will be reached at a distance of $\approx$1100 µm to the SNSPD, since for larger distances the stray light does not show further attenuation and the SNR would decrease.

7.4.4 On-Chip Detection of Emission of a Resonantly-Excited QD

After the response of the WG-SNSPD to photons emitted from a QD ensemble, the response of the detector to the emission of a single resonantly QD was investigated. To choose a QD to resonantly excite, an ensemble of QDs was excited above-band and characterized using the off-chip spectrometer. The resulting spectrum can be seen in fig. 7.26a. The brightest line (at 872.56 nm) was picked and excited resonantly by tuning the excitation laser to the QD wavelength, spectrally shaping it and stabilizing the emission using an above-band stabilization laser at $\lambda = 633$ nm (see section 2.2.1 of chapter 2).

The photo luminescence spectrum of the single resonantly-excited QD is shown in fig. 7.26b. The laser background is characterized by a switching off of the QD excitation using optical gating by the stabilization laser [128]. The SNR is hereby calculated as the relation of the QD intensity to the background intensity $\frac{I_{\mathrm{QD}}}{I_{\mathrm{Backg}}}$ to 9. This SNR is sufficient for an experiment on a single-photon level with detection off-chip.

The count rates on the WG-SNSPD in dependence on the relative bias current is shown for a single resonantly-excited QD in fig. 7.27a. The stray light counts were measured by slightly detuning the excitation laser from the resonance wavelength and subtracting dark counts. The QD counts were measured under resonant pulsed excitation from the top and are shown after subtraction of dark and stray light counts. The dark count rate shows traces of thermal count (section 7.3.2). The number of measured stray light counts is higher than the number of QD

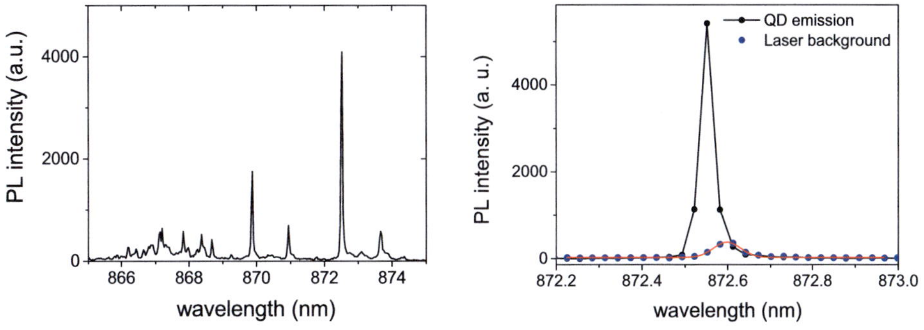

(a) QD ensemble

(b) single QD

Figure 7.26: (a) Photo-luminescence spectrum of a QD ensemble in the WG for above-band excitation. (b) Photo-luminescence spectrum of a single resonantly-excited QD in the WG (black) at 872.56 nm. The laser background on the spectrometer is shown in blue and was fitted with a Gaussian (red) function.

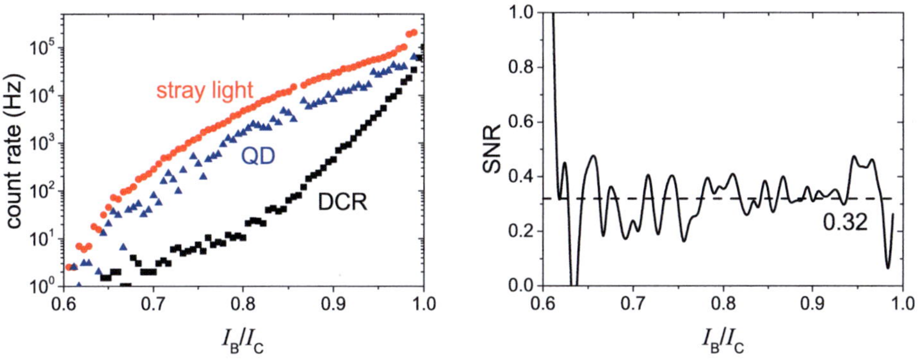

(a) SNSPD count rate

(b) SNR

Figure 7.27: Bias dependence of the SNSPD count rate (a) and SNR (b). The bias depencence of the count rate for dark counts in black, stray light counts in red and QD counts in blue.

counts and significantly higher than dark counts. The SNR is shown in dependence of the SNSPD bias current in fig. 7.27b. It can be seen that stray light is the most dominant source for noise counts. Since stray light is registered on the SNSPD with the same DE as QD photons, the SNR scatters around ≈0.32 independent of the bias current. A SNR of 0.32 is not sufficient for a successful measurement on the single-photon level and the SNR needs to be improved further.

7.4.5 On-Chip Decay Time Characterization

To evaluate the possibility of time gating to improve the SNR for the experiment, the timing characteristic of the resonantly-excited QD is evaluated on-chip. Time gating can be used to cut out laser stray light during pulsed-resonant excitation by utilizing the different time

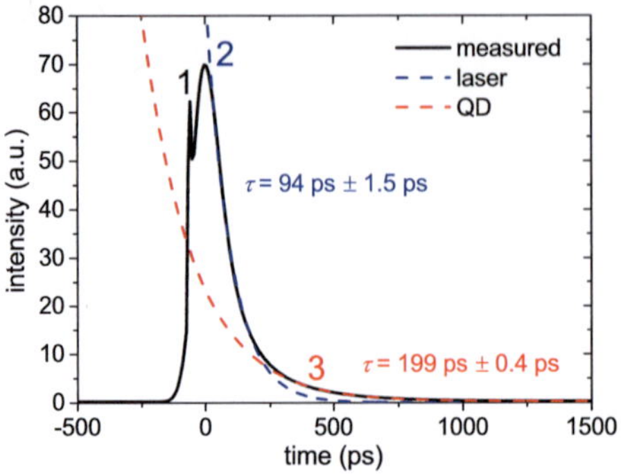

Figure 7.28: TCSPC characteristic of a resonantly-excited QD. The dashed lines are exponential fits to the data.

characteristics of the laser pulse in comparison to the emission of the QD. The laser pulse has a length in the order of several picoseconds while the QD emits its photon over a time in the order of several hundred picoseconds. The TCSPC measurement was performed by recording WG-SNSPD pulses in relation to the trigger signal of the excitation laser as a time reference.

The time characteristics is shown in fig. 7.28. The first peak (1) and the second peak (2) are caused by laser stray light while the tail (3) is caused by QD light. Since peak (1) is very sharp and too short for shaped laser light, it can only be explained only be explained by stray light from the excitation laser before spectral shaping that enters cryostat from free space due to insufficient shielding. After pulse shaping, the pulse length of the laser is significantly longer and a decay time of 94.0 ± 1.5 ps is extracted out of an exponential fit to decay of the second peak (2). The decay time of the QD is extracted by an exponential fit of the tail (3). The fit reveals a decay constant of 199.0 ± 0.4 ps. This time fits very well to the expected decay time of the excited QDs and can be attributed to QD emission. A SNR for the QD emission of 0.46 can be calculated from the area of the fit to the QD decay behavior compared to the total area of the measured time characteristic. This is too low for an on-chip measurement on the single-photon level. By using time gating the laser peak can be cut out by utilizing its faster decay. If only counts in a time window from 250 to 1500 ps are investigated, the SNR increases to 2.6. However, only 28 % of the total QD emission are used for experiment. By further decreasing the time window from 500 to 1500 ps, the SNR can be increased to 13 with only 9 % of the total QD emission remaining. This would make fully on-chip measurements on a single-photon level possible. However, since the main number of QD counts is also cut, the measurement time needs to be increased significantly. Furthermore, the peak (1) can be removed by improving the optical shielding of the cryostat.

7.4.6 Discussion

The on-chip emission and detection of photons was successfully demonstrated for a QD ensemble excited above-band and a resonantly-excited QD. The measured brightness was investigated along with noise counts to evaluate the SNR for a fully integrated experiment.

Since the measured SNR of 0.46 for resonant excitation is too low for the demonstration of a quantum-photonic circuit on a single-photon level, several actions have to be undertaken for improvement. Whilst measurements using pulsed excitation are possible by filtering out the excitation laser using time gating, time gating has several disadvantages: a large amount of QD photons are filtered out and it does not work for cw-resonant excitation. However, several measures were found that improve the SNR: for a WG attenuation below $3.5 \pm 0.2\,\mathrm{dB/mm}$ the optimal excitation distance to the SNSPD is $\approx 1100\,\mu\mathrm{m}$, for a higher WG attenuation the QD excitation should be performed as close as possible to the taper. At this distance, using a cover layer on the SNSPD reduces the stray light counts by a factor of 2.6 since direct stray light is blocked from the SNSPD. A further reduction by a factor of 3 to 6 at this distance is achieved by the use of a dual-side polished substrate. Further samples were made with sputtered silicon and germanium as absorber on the back side of the substrate but could not be characterized. Silicon was sputtered on the backside before detector fabrication, but the adhesion of sputtered silicon to GaAs was too low and did not survive the fabrication process. Gallium was sputtered after sample fabrication on a sample containing QDs and WG-SNSPD. However, the WGs were damaged and the chip could not be characterized. No stray light in the WG was measured on-chip. So for a covered SNSPD, all stray light reached the detector through the substrate. Therefore, a better shielding of the detector can be reached by using large covers on the full chip to prevent stray light photons from entering the substrate. At the level of the 20 K thermal shield, the implementation of an optical high pass filters would remove thermal counts from the system with the drawback of making resonant excitation more challenging.

7.5 Fully On-Chip Integrated Quantum Circuit

Whilst the photon emission and detection on the same chip utilizing WGs and SNSPDs have been demonstrated in the previous section, the integration of a photonic logical element and the demonstration of the operation on a single-photon level is still missing for the demonstration of a fully integrated quantum-photonic circuit. To demonstrate the opration on a single-photon level, a full Hanbury-Brown and Twiss (HBT) setup is monolithically integrated on-chip. In this section, the design of such a setup on-chip is introduced. All individual elements of the setup are characterized and, finally, the first on-chip Hanbury-Brown Twiss measurement is carried out and the results are evaluated. Results shown in this chapter are published in [SSR$^+$18].

7.5.1 Chip Design

An illustration of a fully integrated HBT setup is shown in fig. 7.29. Two single-photon detectors need to be integrated on a WG together with a BS with a splitting ratio close to 50/50 along with a WG-integrated single-photon source. Al/AlN covers are deterministically placed on top of the detector (not shown) and to the sides of the WG to block direct stray light from the detector and stray light from entering into the substrate. The design consists of

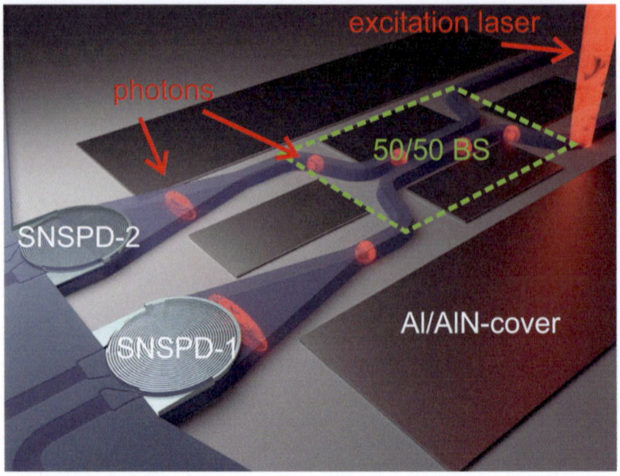

Figure 7.29: Illustration of a fully integrated Hanbury-Brown Twiss experiment with two SNSPDs, a 50/50 beam splitter (BS) and the excitation of a waveguide-integrated quantum dot. Reprinted with permission from [SSR+18]. Copyright 2018 American Chemical Society.

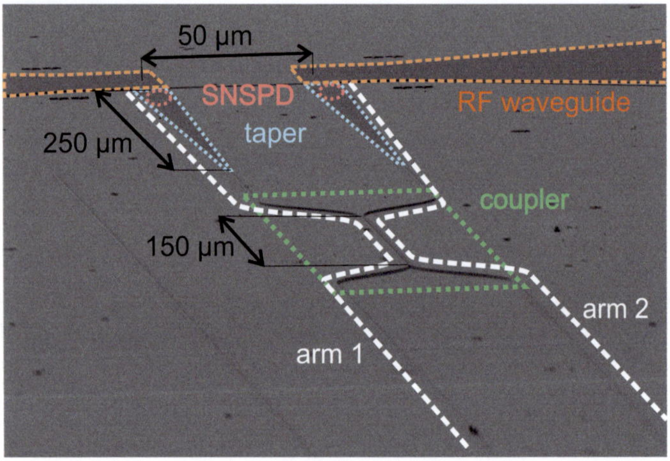

Figure 7.30: Tilted SEM image of a fully on-chip HBT setup. The RF waveguides to readout and bias the detectors are indicated in orange. Positions of the SNSPDs are shown in red, of the taper in blue and of the coupler in green.

two single-mode GaAs WGs at a distance of 50 µm with embedded self-assembled InGaAs quantum dots. A waveguide-based directional coupler used as a beam splitter. After the beam splitter, the WG is tapered to a width that allows the integration of two spiral type SNSPDs. A SEM image of a chip without Al/AlN covers is shown in fig. 7.30.

For the full experiment a single quantum dot is excited resonantly by a laser in one of the WG arms and emits photons that propagate in the WG in both WG directions. In one direction, the photons travel 381 µm in the waveguide to the end of the chip and are coupled into free space at

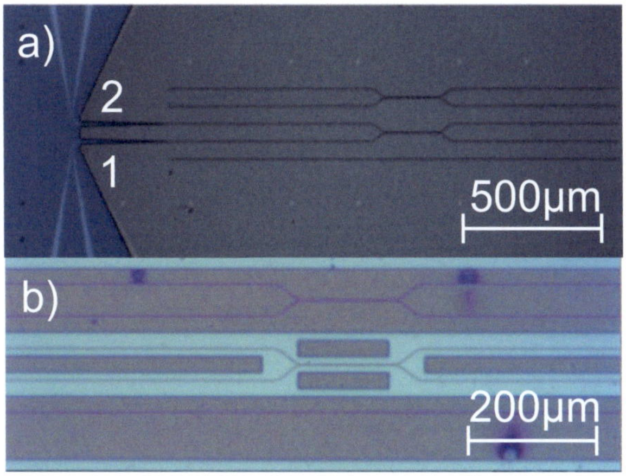

Figure 7.31: a) Optical image of the sample from the top before the fabrication of covers. The coupler formed by the two center WGs is connected to the detector. The coupler above and the WG below are test structures. b) Optical image of the coupler region after cover fabrication. The cover fully covers the test structures.

the cleaved sample edge. In the other direction, the photons travel 1082 μm in the waveguide and propagate across the 50:50 splitter to the two independent WG SNSPDs connected to the output arms of the beam-splitter, SNSPD-1 and SNSPD-2. The beam-splitter has a length of 150 μm and a gap of 95 ± 5 nm. After the BS the photons are guided in the 580 ± 10 nm wide waveguide for another 500 μm until they reach the taper. The length of this WG allows to cleave the sample between the SNSPD and the BS to investigate the SNSPDs and the BS independent from each other. The used taper is a linear taper at a length of 250 μm. Using the taper the WG is broadened from a width of 580 ± 10 nm to a width of 10 μm to embed the 7.6 μm wide SNSPD. The SNSPDs were structured out of a 4.2 nm thick NbN film on a 10 nm thick AlN buffer layer. The NbN film has a critical temperature of 10.4 K. The SNSPDs were fabricated in the double spiral design introduced in chapter 4. The nanowires have a total length of 280 μm, a width of 115 ± 5 nm and a gap of 85 ± 5 nm between wires. The WGs, including the BS, were patterned on the detector chip using electron beam lithography with 55 nm thick HSQ resist and an etching process performed via inductively coupled plasma reactive ion etching in Argon + $SiCl_4$ atmosphere. After etching, the HSQ and the AlN buffer layer remained on the WG. A top view optical image of the WGs including taper and splitters is shown in fig. 7.31a. The additional WGs seen above and below the WGs connected to the SNSDPs are test structures.

To improve the SNR and reduce stray-light counts of the SNSPDs, the design of the sample was fitted accordingly. The SNSPDs were covered in a deterministic way using an Al/AlN bi-layer (fig. 7.32). The AlN below the Al layer is used as a transparent spacing layer to prevent absorption of WG-guided photons in the Al cover. The SNSPD covers have a width of 9 μm and a length of 10 μm on the WG. Off the WG the cover is enlarged by a rectangular cover with a width of 10 μm and a length of 70 μm. The enlarged cover further prevents photons from entering into the substrate in close proximity of the detector. It was not further enlarged in order to prevent an influence of the cover on the RF readout of the detector. Furthermore, the

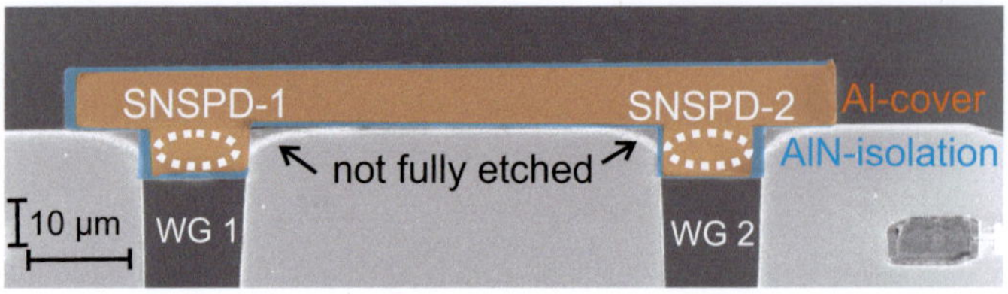

Figure 7.32: Tilted false-color SEM image of the covered SNSPDs. The Al cover is colorized in orange and the AlN isolation underneath in blue.

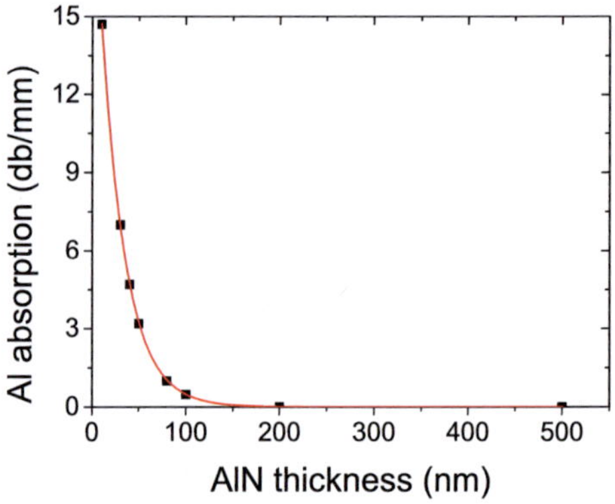

Figure 7.33: Absorption of photons in the Al-cover in dependence of the AlN thickness.

4.2 nm thick NbN layer of the RF line is semitransparent and absorbs 34 to 38 % of photons, preventing them from entering the substrate. The absorption of photons in the Al-top layer was simulated for the 10 µm wide WG in the detector region by Stefan Hepp. The simulation results are shown in fig. 7.33. The simulation was done including the 50 nm transparent SiO_x layer that is left on the WG after patterning. At an AlN thickness of 50 nm only 0.32 % of WG-guided photons are absorbed in the Al/AlN bi-layer over the length of the cover of 10 µm on top of the detector. Hence, an AlN thickness of 50 nm is sufficient. The transition from the taper to the not structured readout area of the WG is not etched perfectly due to the proximity effect during EBL (fig. 7.32). However, this should not be of much concern since it only influences the last 5 µm of the WG while the rest of the WG is homogenous.

In addition, to detector covers, Al/AlN covers are placed close to the waveguide structure and in between the coupler arms to prevent the entry of thermal photons and photons from the excitation laser and the stabilization laser into the substrate. The covered chip is shown in comparison to the uncovered chip in fig. 7.31a, b. The covers consist of two rectangular covers at a distance of 25 µm to the WG and BS structure over the full length of the etched WGs, together with two rectangular covers next to the BS on both sides at a distance of 10 µm to the coupler and a length of 130 µm and two covers in between the two waveguides at a distance of 10 µm to the WGs over the full length of the etched WGs. The distance to the WG helps to reach a high fabrication yield without liftoff issues. The test WGs were overgrown during cover fabrication.

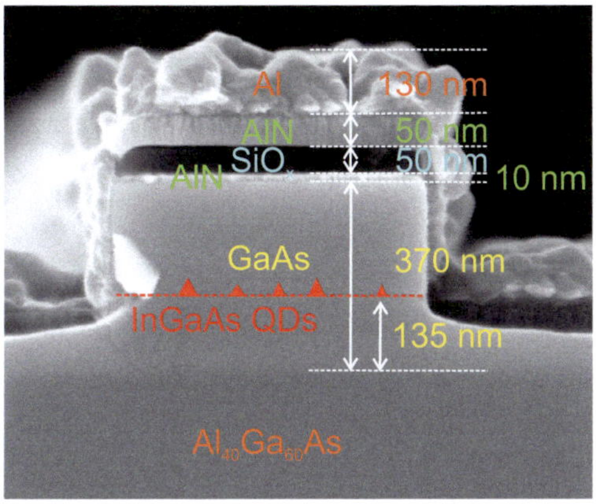

Figure 7.34: SEM image of the layer structure of the sample as seen for the cleaved faced of one of the fully covered test WGs.

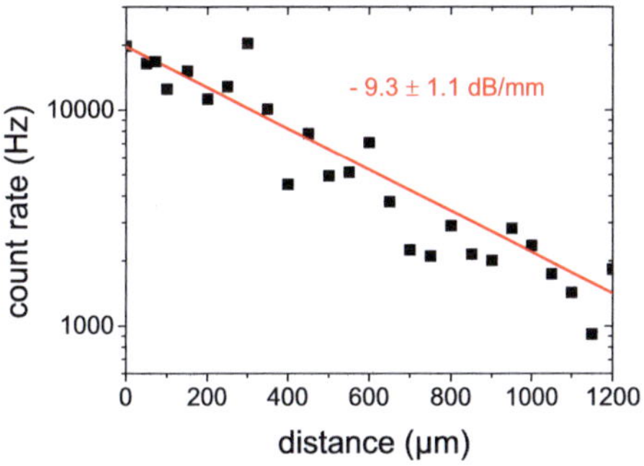

Figure 7.35: Count rate as a function of the excitation distance to the cleaved WG end. The red line is an exponential fit to the measured data and gives the absorption of the WG.

The resulting layer structure can be seen in the cross-section of the cleaved WG face of an overgrown test WG (fig. 7.34). The WGs are fabricated on top of a 2 µm thick $Al_{0.4}Ga_{0.6}As$ layer out of a 370 nm thick GaAs-layer with integrated QDs at a height of 135 nm. The WG is etched to a depth of 320 nm. On top of the WG a 10 nm thick AlN buffer layer followed by a 50 nm thick SiO_x layer is present. The SiO_x is a result of the WG fabrication with HSQ resist and remains on the WG, SNSPDs and RF-readout lines. On the top, the Al/AlN bi-layer that is used as a cover-mirror is visible. The AlN isolation layer has a thickness of 50 nm and shows a smooth structure followed by a granular Al film with a thickness of 130 ± 15 nm.

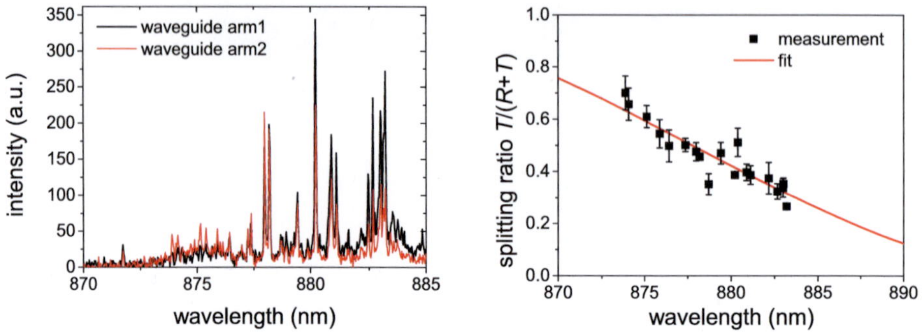

(a) Spectral emission for both WG arms

(b) Wavelength dependent splitting ratio

Figure 7.36: Emission spectrum of an excited QD ensemble and the corresponding wavelength-dependent splitting ratio. The red fit to (b) is a fit to the data with a $\sin^2$ function. Figure (b) was reprinted with permission from [SSR+18]. Copyright 2018 American Chemical Society.

7.5.2 Characterization of the Integrated Waveguide and Beam Splitter

Absorption of the Waveguide

The WG is characterized using above-band excitation of WG-integrated QDs at $\lambda = 800$ nm. The laser is focused on the WG and moved along whilst measuring the count rate off-chip using an APD fig. 7.35. The WG attenuation was extracted using an exponential fit. The fit reveals an attenuation of 9.3 ± 1.1 dB/mm. This attenuation is higher than for state of the art WGs, for which attenuation levels of 2.6 ± 0.6 dB/mm were achieved [67]. Possible causes for this higher attenuation are the surface roughness of the WG sidewalls and the remaining of the AlN buffer layer on top of the WG.

Splitting Ratio of the Beam Splitter

The splitting ratio of the BS was characterized by excitation of a QD ensemble in WG arm 1 (see fig. 7.30 on page 146) on the SNSPD side of the BS and measuring the emitted spectrum for both WG arms using the off-chip spectrometer. The emission spectrum is shown in fig. 7.36a. Sharp spectral lines can be seen. Each of these spectral lines represents one QD. To determine the wavelength-dependent splitting ratio, 18 lines out of the spectrum were fitted by a Gaussian function for both WG arms. The splitting ratio was then calculated by comparing the area of those Gaussian fits and is shown in fig. 7.36. Photons that cross the BS and exit the WG at arm 2 are regarded as transmitted photons T, while photons that exit the BS at arm 1 are regarded as reflected photons R. The splitting ratio is then calculated as:

$$\mathrm{SR} = \frac{T}{T + R}. \tag{7.6}$$

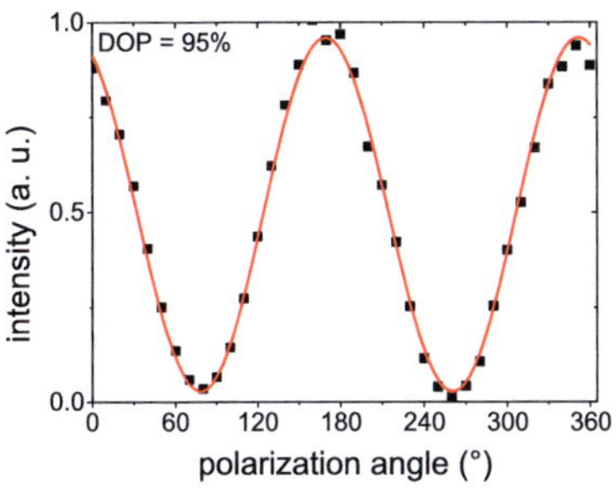

Figure 7.37: Polarization measurement of the emitted QD light from the sample in free space. The red line is a fit of the intensity with a sinus.

The splitting ratio (SR) for any wavelength can be extracted by the use of a sin^2 fit in framework of the super-mode theory. Further details on the super-mode theory for the splitting ratio can be found in [110]. The QD investigated under pulsed-resonant excitation in the following sections emits at a wavelength of approximately 876.1 nm. The splitting ratio at this wavelength is $\frac{56.7}{43.3} \pm 4.4\%$, which is close enough to 50/50 to enable a HBT measurement.

Degree of Polarization

The polarization dependence of the emitted photons was investigated under resonant excitation (for details see following section). For experiments utilizing indistinguishable photons it is important to have an identical polarization of both photons. The single-mode WG design supports two polarization modes in principle: the fundamental TE, as well as the fundamental TM mode which are linear polarized modes with a polarization perpendicular to each other. However, connected to the QD growth, the dipole moment of the QD is typically aligned to the growth plane, with a TM mode perpendicular to the surface plane, which cannot be excited by top excitation. Consequently, only a propagation of the TE mode in the WG is expected [110]. The intensity was measured off-chip for different polarization angles by using a linear polarization filter in the free space optical path. The result is displayed in fig. 7.37 and fitted by a sinus. Overall, a degree of polarization for the emitted light of 95 % could be measured. This shows that only the fundamental TE-mode travels in the WG and most of the emitted photons share the same polarization. The presence of only a single polarization mode proves that the WG is indeed single-mode at the investigated QD wavelength.

7.5.3 Off-chip Characterization of a Single Quantum Dot

In order to investigate whether the quantum dot used is suitable for an on-chip experiment and to provide a reference measurement for the on-chip experiment, the intensity, the emission spectrum and the single-photon purity of the quantum dot were determined off-chip. The off-chip characterization was performed for pulsed as well as cw-resonant excitation. The quantum dot was chosen by excitation of a quantum dot ensemble in the WG on the cleaved

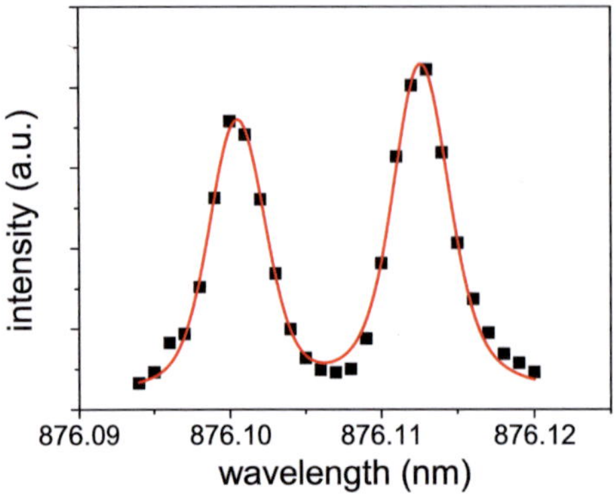

(a) Resonance scan

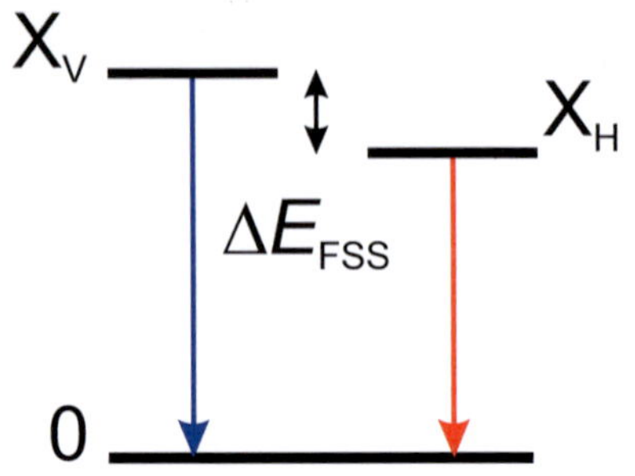

(b) Model of a QD fine-structure splitting (FSS)

Figure 7.38: (a) Intensity of emission for resonant excitation at different excitation wavelengths. Two excitation peaks are visible, which indicates a fine-structure splitting (b). The solid red line of is a fit to the measured intensity with a double-Voigt profile. (b) Model of the present fine structure splitting of the QD energy level with two dipoles with an exciton level in horizontal X_H and one in vertical orientation X_V.

side relative to the BS by searching for a bright sharp line in the emission spectrum. A sharp emission line was found at a wavelength of approximately 876.1 nm and further investigated using resonant excitation. For CW excitation the laser was tuned to the wavelength of the QD with a spectral width of the emission of 500 MHz. During pulsed-resonant excitation the laser is set to the wavelength of the QD and has a spectral width of FWHM = 100 GHz. Using a monochromator, the pulse is spectrally shaped to a FWHM of 15 GHz which results in a pulse length of FWHM = 35 ps. The polarization of the excitation laser was chosen parallel to the WG to suppress the coupling of the laser light into the WG [68]. In addition to the pump laser, a low intensity stabilization laser at a wavelength of 633 nm was used for optical gating to reduce charge fluctuations (see: section 2.2.1 in chapter 2). The stabilization laser has an intensity of >1 ‰ compared to the pump laser. The QD emission is guided over a length of approximately 400 µm in the WG, coupled out through the cleaved sample facet and send to the off-chip spectrometer or the off-chip HBT setup.

Resonance Scan of the Integrated QD

The resonance spectrum of the QD was investigated by an adjustment of the excitation wavelength with the pulse shaper. The measurement (fig. 7.38a) was done by performing a sweep of the excitation wavelength from 876.095 to 876.012 nm close to the QDs energy level and a recording of the QDs intensity response. In the response, two distinct intensity peaks are visible: a first intensity peak at a wavelength of 876.101 nm with a spectral width of FWHM = 1.78 GHz and a second intensity peak at a wavelength of 876.113 nm with a spectral width of FWHM = 1.72 GHz. Each of the peaks is best fitted by a Voigt profile and both can be fitted together by a double Voigt distribution, which is a convolution of a Gaussian and Lorentzian function. The Lorentzian part represents the linewidth of the QD

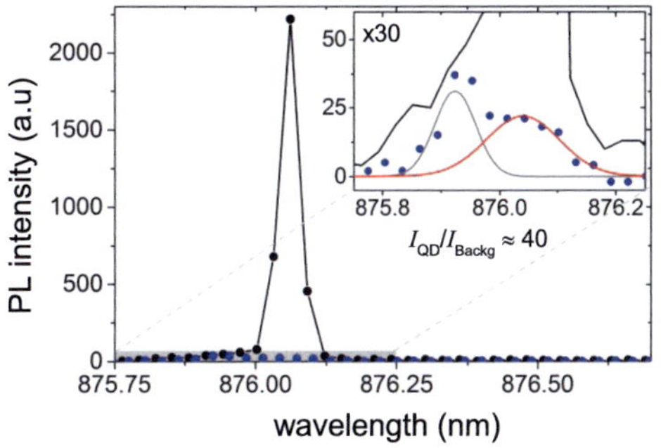

Figure 7.39: Off-chip photo luminescence (PL) spectrum of the quantum dot under pulsed-resonant excitation. The black measurement points represent the PL intensity under resonant excitation. The blue points are measurement with the excitation laser detuned from the resonant wavelength. The inset is a magnification of the measured data at the foot of the visible peak. The gray line is a Gaussian fit to the laser background and the red line is a Gaussian fit to the QD emission caused by the stabilization laser. Reprinted with permission from [SSR+18]. Copyright 2018 American Chemical Society.

and the Gaussian part the spectral jitter of the QD. The presence of two peaks is a sign for a fine structure splitting of the energy levels (fig. 7.38b) with a $\Delta E = 4.60\,\text{GHz}$ or for the presence of a second excited QD with a similar energy level. A fine structure splitting is caused by an asymmetry in the QD. The spatial shape of the QD can be considered to be more elliptical than round. The energy level of the QD is a sum of two different dipoles orthogonal to each other with slightly different lengths with one exciton level in horizontal X_H and one in vertical orientation X_V [221]. The fine structure splitting and respective linewidths for the investigated QD is comparable to QDs in bulk material [219] and no degradation of the optical quality due to the fabrication process can be observed. The absence of a broadening of the linewidths indicates that the QD emission is not affected by surface effects [222], which gives a likely position of the QD close to the middle of the WG [124]. To exclude the excitation of more than one QD, further investigations are performed below.

Emission Spectrum

The photo luminescence (PL) spectrum of the QD emission was investigated and is shown for pulsed-resonant excitation in (fig. 7.39). Stray light counts were characterized by detuning the pump laser whilst maintaining intensity and are shown as blue measurement points. Stray light counts caused by the pump laser were fitted with the Gaussian shown in gray and QD emission caused by the excitation laser was fitted with the Gaussian shown in red. A SNR of 40 is calculated from of the individual areas. A SNR of 40 enables a theoretical $g^{(2)}(0) = 0.048$ in case the QD is an ideal single-photon source. Due to a smaller line width for CW-resonant excitation, the QD excitation is more efficient and the SNR increases to 80.

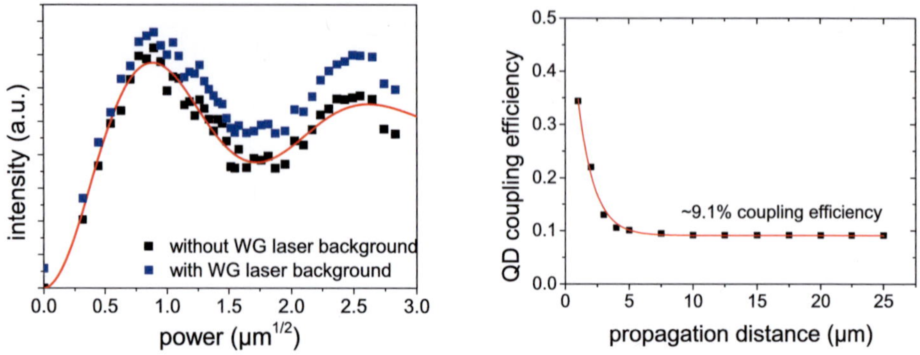

(a) emission power dependence

(b) QD to WG coupling

Figure 7.40: (a)Intensity of QD emission as a function on the square root of excitation power including (blue) and excluding stray light (black) Reprinted with permission from [SSR+18]. Copyright 2018 American Chemical Society. The red line is a fit of the optical Bloch equations to the data. (b) Coupling efficiency of QD photons to the WG.

Power Dependence and Emission Efficiency of the Quantum Dot

The emission intensity of the QD was characterized off-chip using pulsed-resonant laser excitation (fig. 7.40a). Excitation of the QD at the first maximum, which is connected to the highest excitation efficiency, is called π-pulse excitation. A fit with the optical Bloch equations reveals clear Rabi oscillations of the QD. This proves the presence of a coherent excitation process. Out of the fit a state preparation of 70 ± 1 % can be extracted. The state preparation is the transition moment of the Rabi oscillations and gives the probability of an excitation pulse leading to the emission of a QD photon. An exact description of the fitting can be found in [124]. For the following experiments the QD was excited at 90 % of the π-pulse. The coupling efficiency of QD photons to the WG was simulated using a finite element time domain simulation by Stefan Hepp in dependence of the propagation distance in the WG. Out of the simulation (fig. 7.40b) a coupling efficiency of 9.1 % for the TE mode can be extracted for a propagation length >5 μm. Only 50 % of emitted photons couple to the TE-mode and can propagate inside the WG. The traveling distance from the QD to the cleaved WG edge is 381 μm. With the previously measured WG attenuation, 44.2 ± 4.7 % of photons are transmitted. From the photons that couple out of the WG-facet 20.4 ± 1.8 % are collected [110]. The combined efficiency for a collected photon to go through the free space optical path and to trigger an APD count was determined to be 4.13 ± 0.49 % [124]. With the PRF of the laser of 76 MHz, the expected count rate on the APDs is calculated to 8113 ± 3628 Hz. This fits very well to a measured count rate on the APDs of ≈10 kHz within the accuracy of the measurement.

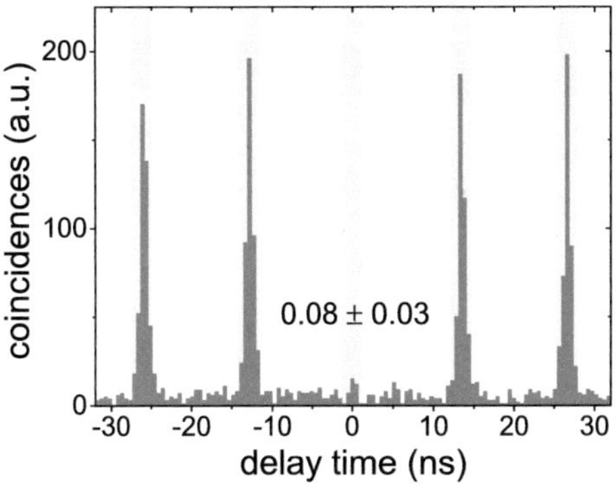

Figure 7.41: Second order correlation for QD emission under pulsed-resonant excitation characterized using the off-chip setup. Reprinted with permission from [SSR+18]. Copyright 2018 American Chemical Society.

2nd Order Correlation

The quality of single-photon emission was characterized utilizing the off-chip HBT setup with APDs to evaluate the single-photon emission of the QD with a conventional setup. To compare the device performance between off- and on-chip detection, no spectral filtering was used at the detection side of the experiment. The QD was excited at 90 % of the π pulse. Under pulsed-resonant excitation coincidence peaks can be observed with a distance of approximately 13 ns which corresponds to the repetition rate of the excitation laser of 76 MHz. The number of coincidences are clearly suppressed at zero time delay of the detector responses (fig. 7.41). By dividing the average intensity of the peaks at $t \neq 0$ with a width of 2.56 ns by the number of coincidences at zero time delay, a $g^2(0)$ of 0.08 ± 0.03 can be extracted. The non-vanishing part is caused by the remaining small amount of laser stray light, a possible re-excitation of the QD due to pulse length of the laser excitation of 35 ps and an excitation of the QD by the stabilization laser. The emission of the QD is, therefore, proven to be clearly on the single-photon level and the two emission peaks during the performed resonance sweep are caused by the fine structure splitting in the single QD and not by two QDs close to each other. From the similar intensities for both fine structure emissions it can be concluded that the QD polarization is rotated by close to 45° to the WG and, therefore, also to the laser polarization.

7.5.4 Detector Characterization

In order to evaluate the quality of the on-chip detectors, they were investigated using laser light. The SNSPDs were characterized at a temperature of 4 K in the free-space setup using laser emission from the top at a wavelength of 870 nm. Since the detectors cannot be directly illuminated because of the Al/AlN cover, the laser was centered between both SNSPDs at a distance of 25 µm to the SNSPDs at a position without cover. Stray light reflected at the backside of the substrate can be seen by the SNSPDs and is used for characterization. The

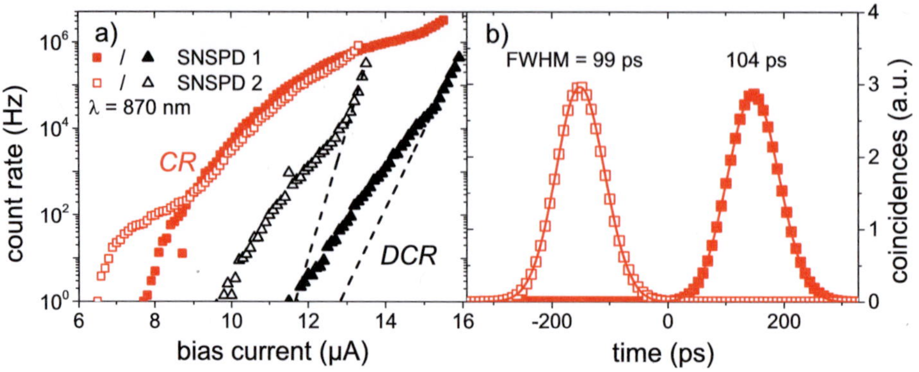

Figure 7.42: a) Count rates as a function of the bias current of the detectors. Red squares show the count rates (CRs) by excitation of the detectors with laser stray light. Black triangles show the DCR of the SNSPDs. Results obtained on SNSPD-1 are shown in filled symbols and on SNSPD-2 are in open symbols. Dashed lines show an exponential fit to the exponential trend of the DCR just before the SNSPD's critical current. b) Time response of the detected signal for both SNSPDs in correlation to the laser trigger, in filled symbols for SNSPD-1 and in open symbols for SNSPD-2. The corresponding continuous lines are Gaussian fits to the measured data. Reprinted with permission from [SSR⁺18]. Copyright 2018 American Chemical Society.

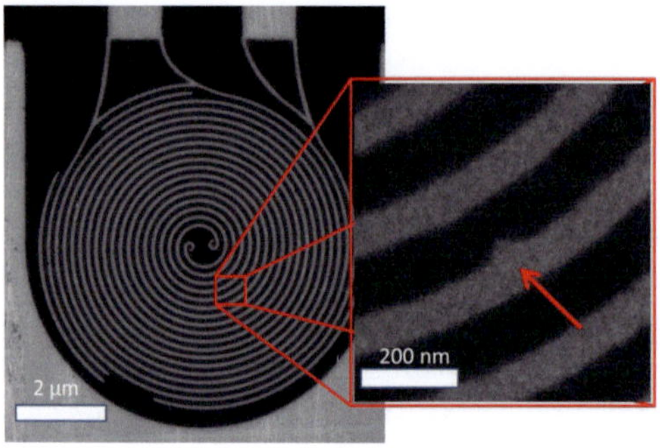

Figure 7.43: SEM image of SNSPD-2, the exact position of the defect is shown in the inset.

SNSPDs were investigated regarding their bias dependence (fig. 7.42a) and timing characteristics (fig. 7.42b).

Bias Dependence of the On-Chip SNSPDs

Between a bias current of 9 to 12 µA the count rate dependencies coincide with each other. SNSPD-1 has a critical current of 15.9 µA and SNSPD-2 a critical current 13.5 µA. This limited critical current of SNSPD-2 stems from a 50 nm long single constriction. In the constriction, the width of the nanowire is reduced from a width of 115 ± 5 nm to a width of 85 ± 5 nm. Hence, SNSPD-2 is limited in count rate and corresponding detection efficiency.

The bias dependence of SNSPD-2 shows a further influence of the constriction. At a bias current of 6.3 μA, a strong increase in count rate can be seen which is caused by the constriction becoming sensitive to photons. At approximately 9 μA, the count rate shows a second strong increase since the full nanowire is sufficiently biased to get sensitive for single photons and fully contributes to the measured count rate. This behavior is a clear sign for a constriction defect in the detector [223] and is confirmed by the SEM image (fig. 7.43). Caused by the indirect illumination and the unknown photon distribution of the stray light on the detector the laser cannot directly be used for the characterization of the DE. Hence, the detection efficiency (DE) is characterized using a single resonantly-excited QD on-chip. This has the further advantage that, by calculating the DE out of the QD emission, the DE for the experiment can be extracted. The laser excites the QD at a repetition rate of 76 MHz. With the power-dependent measurements in section 7.5.3 giving a state preparation of 70 ± 1 %, the calculated coupling efficiency of the QD of 9.1 %, the propagation losses in the WG (section 7.5.2) over a distance of 1082 μm and the splitting ratio of the WG of $\frac{56.7}{43.3}$ ± 4.4% (section 7.5.2), the expected number of photons at the detector position can be calculated analog to the calculation of expected APD count events in section 7.5.3. This results in an estimated number of photons of 132 855 ± 43 975 at SNSPD-1 and of 101 458 ± 34 191 at SNSPD-2. At a bias current of 13.5 μA/11.9 μA (SNSPD-1/SNSPD-2), the DE was extracted to be 21.8 ± 7.2 %/1.8 ± 0.6 %. Out of the bias dependence of the detectors the detection efficiency for a higher bias level can be calculated to 47.5 ± 15.7 %/11.2 ± 3.8 % at a bias current of 15.2 μA/13.0 μA.

Dark Count Rate

The dark count rates of the SNSPDs show clear signs of thermal noise counts which can be attributed to the free-space setup 7.3.2. When decreasing the bias currents below 15.2 μA for SNSPD-1 and 13 μA for SNSPD-2 respectively, it deviates from an exponential bias dependence, which would be expected for dark counts intrinsic to a SNSPD [146]. The SNSPD-2 suffers from a high number of dark counts caused by the higher current density in the constriction.

Instrumental response function

To evaluate the timing characteristics of the detecion setup, the instrumental response function (IRF) was characterized: The IRF in fig. 7.42(b) was measured for the full on-chip detection setup for a pulsed laser excitation at a wavelength of 890 nm without the use of the pulse shaper at a pulse width of 3.3 ps. It is equivalent to the system timing jitter of the full setup. The bias current for both detectors was set to 12 μA. This bias level was chosen to operate both SNSPDs at an identical absolute bias condition without regarding the defect in SNSPD-2. $\text{IRF}_{\text{single}}$ is the IRF for the correlation of a single SNSPD with the laser trigger as used for a TCSPC measurement. It is the convolution of the timing characteristics of the individual components in the free-space setup: The timing variations of the corresponding electrical trigger signal τ_{trigger}; the timing jitter of the corresponding detector τ_{SNSPD}; the timing characteristic of the amplifiers and electronic cables τ_{amp}; the timing resolution of the used TCSPC electronic $\tau_{\text{HydraHarp}}$. A Gaussian fit reveals a full width half maximum (FWHM) value of 104 ps for the detection setup with SNSPD-1 and 99 ps for the detection setup with SNSPD-2 at a bias current of 12 µA. Due to a short pulse-duration of approximately 3.3 ps, the influence of the unshaped excitation laser pulses can be neglected.

$$\text{IRF}_{\text{single}} = (\tau^2_{\text{trigger}} + \tau^2_{\text{SNSPD}} + \tau^2_{\text{amp}} + \tau^2_{\text{HydraHarp}})^{1/2} = 101 \pm 3\ ps, \tag{7.7}$$

$\text{IRF}_{\text{System}}$ is the IRF for a correlation experiment, time correlating both detector signals with each other. It is the convolution of the timing accuracy of the corresponding detectors $\tau_{\text{SNSPD-1}}$ and $\tau_{\text{SNSPD-2}}$, and the timing accuracy of the amplifiers and electronic cables τ_{amp} as well as the used TCSPC electronic $\tau_{\text{HydraHarp}}$ for each individual readout channel. The $\text{IRF}_{\text{System}}$ was determined to have a value of 130 ps (FWHM). Since both detectors trigger on the same laser pulse, the optical jitter of the pulse can be neglected. In addition, since the measurement characterizes the two detectors in time relative to each other, no external jitter has to be considered.

$$IRF_{System} = (\tau^2_{\text{SNSPD-1}} + \tau^2_{\text{SNSPD-2}} + 2 * \tau^2_{amp} + 2 * \tau^2_{\text{HydraHarp}})^{1/2} = 130\ ps. \tag{7.8}$$

7.5.5 On-Chip Emission and Detection

Above-Band Excitation of a QD Ensemble

In preparation for the final experiment, first tests of the on-chip detection of QD light were done using a full QD ensemble. Therefore the experiment introduced in section 7.4.2 was performed. The laser was adjusted to an above-band wavelength of 800 nm for excitation. The laser was focused onto the sample surface and driven perpendicular over the cover and the waveguide on the cleaved side of the BS, as seen in section 7.4.2, from $x = -175$ to 175 µm. The zero of the x-axis is hereby defined as the middle of both WG arms. The experiment is performed at a distance of the excitation laser of ≈1100 µm to the SNSPD. Both SNSPDs were adjusted to a bias current of 12 µA for a comparable efficiency. By driving over the on-chip AlN covers, the efficiency of the covers to prevent photons from entering the substrate can be investigated. The measured data is shown in fig. 7.44.

At an x-value of −175 µm the laser spot position is on the GaAs at the side of the Al/AlN cover and the counts on the detectors resembles the stray light. At $x = -150$ µm the laser drives on the cover and the count rate of the detectors decreases until it reaches a stable level. The stable count rate is in the order of magnitude, albeit slightly higher than the DCR. At $x = 50$ µm the laser drives of the cover and the laser already starts to excite some QDs, so the count rate is slightly increased above the stray light level. For the laser spot getting closer to the middle of the first coupler arm at $x = -25$ µm, the efficiency of QD excitation in the WG increases

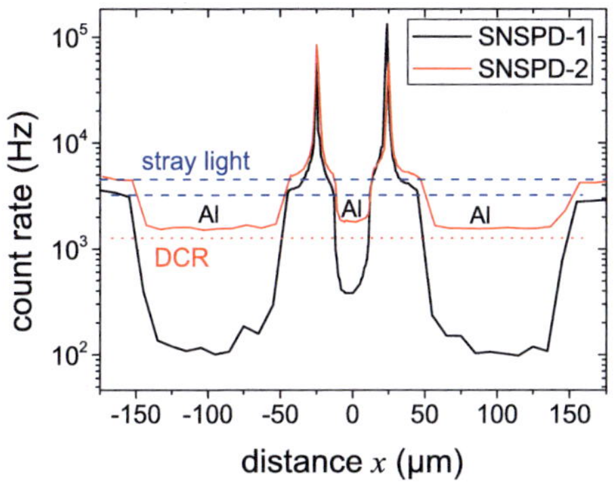

Figure 7.44: Detector response in dependence on the position of the excitation laser relative to the waveguide. The black line shows the position-dependent count rate for SNSPD 1 and the red line for SNSPD-2. The level of stray light is indicated by the dashed blue line. The lower blue line represents the stray light for SNSPD-1, the upper for SNSPD-2. The position of the on-chip Al-cover is indicated. The red dotted line shows the DCR level for SNSPD-2, the DCR level for SNSPD-1 is below the depicted range of the y-axis.

and the count rate reaches a maximum. At $x = -10\,\mu m$, the count rate again decreases when the laser spot is on the Al cover mirror. The general progression of the count rate on laser position is mirrored symmetrical and further follows the already described dependence. The strong suppression of stray light when the excitation laser is on the Al/AlN cover, conclusively demonstrates that most of the stray light is guided through the substrate before reaching the detector. Out of this measurement a splitting ratio of the WG for the full QD ensemble can be extracted. By comparing the heights of the two visible peaks, the splitting ratio calculates to 32 %/68 % for SNSPD-1 and 40 %/60 % for SNSPD-2. A difference in the splitting ratios for each SNSPD is caused by a different brightness for the excited QD ensembles. The measured splitting ratio coincides with the splitting ratio measured off-chip (section 7.5.2). Since a broad band of wavelengths are emitted by the QD ensemble, the exact splitting ratio for the wavelength of the resonantly-excited QD cannot be extracted. The signal-to-noise ratio for excitation in the WG arm 1/arm 2 including dark counts is 18.7/41 for SNSPD-1 and 17.8/11.7 for SNSPD-2. When the dark counts are subtracted, a SNR of 18.7/41 for SNSPD-1 and 25.3/16.5 for SNSPD-2 can be extracted. The accumulated SNR values, calculated by summing up the emission at both sides of the splitter without DCR, are 59.7 for SNSPD-1 and 41.8 for SNSPD-2. The SNR is significantly increased in comparison to the SNR values on the same single-side polished substrate determined for a straight WG (section 7.4.2) from 8 to 41.8. The the increase of the SNR by a factor of 5 can be attributed to the presence of the used on-chip cover layer.

Resonant Excitation of a Single QD

For on-chip measurements, the previously used QD was resonantly-excited to 70 % of the π-pulse. The QD was excited at a distance of 82 µm to the BS on the cleaved side. The

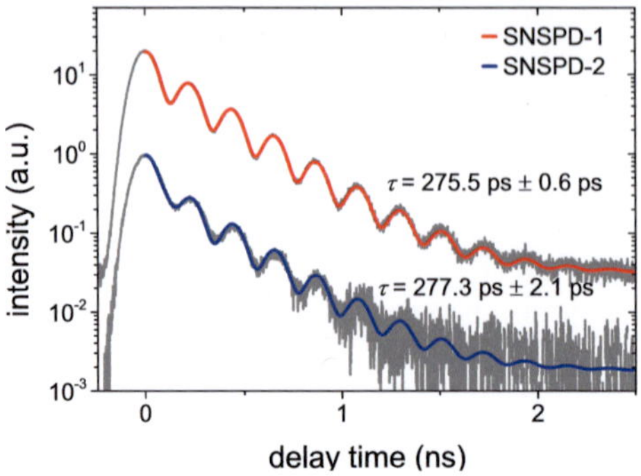

Figure 7.45: On-chip TCSPC measurement of the resonantly excited QD reveal, that the time-dependent intensity follows a mono-exponential decay with overlying oscillations. The red line is a fit to the measured data for SNSPD-1 and the blue line a fit to the measured data for SNSPD-2. For better visibility, measured data points for SNSPD-1 were multiplied by a factor of 10 to separate them from the data of SNSPD-2. Reprinted with permission from [SSR+18]. Copyright 2018 American Chemical Society.

photons are guided through the coupling region of the on-chip beam splitter and are split into the two waveguide arms, where the photons travel 500 µm in the WG, enter the taper and reach one of the two SNSPDs. The SNSPDs were operated at a bias current of 13.5 µA and 11.9 µA since the SNR was lower for higher bias currents due to an increased DCR. At these currents the detectors have a detection efficiency of 21.8 ± 7.2 %/1.8 ± 0.6 % (SNSPD-1/SNSPD-2).
For cw-resonant excitation of the QD on the SNSPD-1/SNSPD-2, an overall count rate of 400 kHz/40 kHz can be measured. In this count rate stray light photons at a rate of 23 kHz/6.5 kHz and dark counts at a rate 0.25 kHz/3.5 kHz are included. The resulting SNR is 16.2/3.0 and 16.4/4.6 when dark counts are subtracted.
For pulsed-resonant excitation of the QD on the SNSPD-1/SNSPD-2 an overall count rate of 36 kHz/7.5 kHz can be measured. The count rate caused by stray light in the pulsed case is 4.8 kHz/1.4 kHz and that caused by dark counts is 0.9 kHz/4.0 kHz. In consequence, a SNR of 5.3/3.0 can be calculated and a SNR of 6.3/4.6 when dark counts are subtracted. The SNR for pulsed-resonant excitation is lower because of a less efficient excitation connected to a larger spectral width of the excitation laser.

On-Chip TCSPC

To evaluate the timing characteristic of the QD and the suitability of the on-chip detectors for on-chip time-correlated experiments, a time-correlated single-photon counting experiment was performed on-chip using pulsed-resonant excitation. The measurement was performed by correlating SNSPD pulses with the trigger signal from the excitation laser. The resulting photon intensity distribution is shown in fig. 7.45.

The time-dependent intensity distribution shows a Gaussian-shaped rise corresponding to the IRF (section 7.5.4) with a subsequent decay. The decay was fitted by an exponential decay with overlying harmonic oscillations.

$$f(t) = A \exp^{(-\frac{t-t_0}{\tau})} \cdot \cos(2\pi f t + \phi), \tag{7.9}$$

where t is the time, t_0 is the time shift of the maximum to 0, A the maximum intensity, f is the oscillation frequency, ϕ is the initial phase and τ is the decay time. t_0, A, f and ϕ are the free fitting parameters. For the fit eq. (7.9) was convoluted with the Gaussian IRF and the pulse duration of the excitation laser pulses (35 ps). Out of the fit a decay time for the QD of 275.5 ± 0.6 ps/277.3 ± 2.1 ps (SNSPD-1/SNSPD-2) was extracted. The decay time should be the same for both detectors and is comparable within the tolerance for the measurement. The decay time is comparable with the decay time for a single QD under resonant excitation measured in a straight waveguide (section 7.4.5). The decay is mono-exponential with superimposed harmonic oscillations. An oscillation frequency of 4.647 ± 0.021 GHz/4.683 ± 0.018 GHz was extracted out of the fit. The oscillations can be explained by the previously seen fine structure splitting by an excitation of both fine structure components. It stems from quantum interference between the horizontal and the vertical component of the exciton and, therefore, a turning of the polarization of emitted photons. The WG acts as a polarization filter for QD emission since only the fundamental TE mode can couple to the WG. These oscillations are considered as quantum beats [224]. The measured frequency corresponds very well to the measured fine structure splitting of 4.60 ± 0.14 GHz by the resonance scan (section 7.5.3). This allows the conclusion that, indeed, a single QD with a splitted fine structure is excited and not two independent dots. Furthermore, the measurement shows that SNSPDs can be used as fast detectors for the on-chip detection of QD emission and allow the measurement of specific properties on short time scales.

7.5.6 Fully On-Chip Hanbury-Brown and Twiss Experiment

After the successful characterization of all on-chip components a fully on-chip Hanbury-Brown and Twiss experiment was performed. A HBT with a $g^{(2)}(0) < 0.5$ proves the operation of all on-chip components together on a single-photon level and proves the operation of the chip on a quantum level. The on-chip second order correlation measurement $g^{(2)}(0)$ is performed similarly to the TCSPC measurement in the previous section. However, the response of one SNSPD is recorded using the response of the second SNSPD as a time reference.

The first HBT measurement was performed using cw-resonant excitation. The small linewidth of the excitation laser allows the excitation of a single fine structure component. The raw data of the cw measurement is shown in fig. 7.46a. At zero time delay a clear anti bunching behavior is visible. At approximately ± 0.5 ns delay a slight bunching behavior is visible. This bunching can be caused by spectral diffusion caused by spectral jumps [225] of the QD line due to electrical fluctuations caused by the excitation laser power [67] and Rabi oscillations [68]. To extract the $g^{(2)}(0)$ value, the fit function includes spectral diffusion and Rabi oscillations:

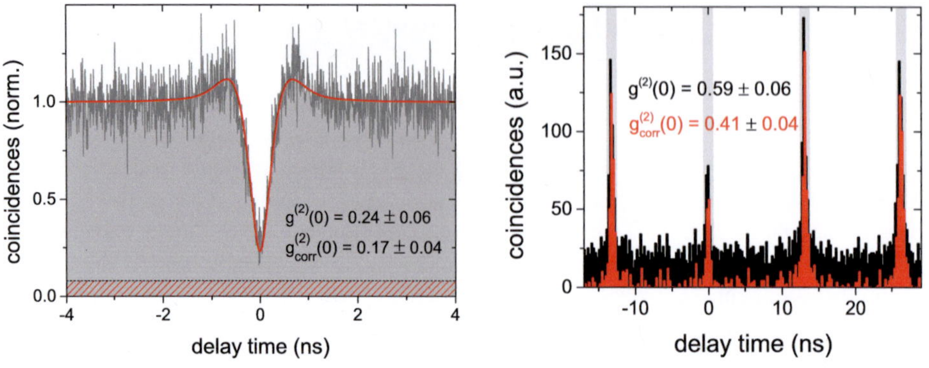

(a) cw-resonant excitation

(b) pulsed-resonant excitation

Figure 7.46: Second order correlation measurement using the full on-chip HBT setup. (a) $g^{(2)}$ of the on-chip QD under cw-resonant excitation. The red line is a fit to the data. The dashed line indicates the dark count level. (b) $g^{(2)}$ of the on-chip QD under pulsed-resonant excitation. The data is shown with (black) and without (red) dark counts, the gray areas indicate the areas for which the $g^{(2)}(0)$ is calculated. Reprinted with permission from [SSR+18]. Copyright 2018 American Chemical Society.

$$g^{(2)}(\tau) = \left(1 - A \cdot \exp^{-\frac{\Gamma_1+\Gamma_2}{2}\cdot|\tau|} \cdot \left(\cos(C\tau) + \frac{\Gamma_1 + \Gamma_2}{2C}\sin(C\cdot|\tau|)\right)\right) \cdot \left(1 + B \cdot \exp^{-\frac{|\tau|}{\tau_b}}\right), \tag{7.10}$$

where A is the maximum of anti-bunching, B the maximum of bunching and τ_b the timescale caused by spectral diffusion. Γ_1 is the radiative decay rate, Γ_2 the damping rate, $C = \sqrt{\Omega^2 - 1/4\cdot(\Gamma_1 - \Gamma_2)^2}$ and Ω the Rabi-frequency. Γ_1 is assumed to be the reciprocal of the measured decay time from the excited QD state (fig. 7.45), and Γ_2 can be calculated out of the excitation power-dependent measurements of the QD intensity I (section 7.5.3) and a measurement of the spectral line width Δ (section 7.5.3) using the theoretical expressions of the emission intensity at zero detuning eq. (7.11) and eq. (7.12) [128]:

$$I \propto \frac{\Omega^2}{\Gamma_1\Gamma_2 + \Omega^2} \propto \frac{P_{\text{laser}}}{P_{\text{laser}} + P_{laser_0}} \tag{7.11}$$

$$\Delta = \frac{h\Gamma_2}{\pi}\cdot\sqrt{1 + \frac{\Omega^2}{\Gamma_1\Gamma_2}} = \frac{h\Gamma_2}{\pi}\cdot\sqrt{1 + \frac{P_{\text{laser}}}{P_{laser_0}}}, \tag{7.12}$$

where P_{laser} is the excitation power of the resonant laser and P_{laser0} is the excitation power at the saturation of the QD emission. With the line width being $\Delta = 1.74\,\text{GHz}$ and the power

$P_{\text{laser}} = 0.43 \cdot P_{\text{laser0}}$ we can calculate $\Gamma_1 = 3.63/\text{ns}$, $\Gamma_2 = 4.52/\text{ns}$ and $\Omega = 2.66/\text{ns}$. The remaining fit parameters are A, B and τ_b. When taking into account the IRF of the used setup of 130 ps, a $g^{(2)}(0)$ value of 0.24 ± 0.06 can be extracted by a deconvolution of the data. If the IRF is not taken into account, a $g^{(2)}(0)$ value of 0.27 ± 0.06 can be extracted. The calculated function has a good conformity to the measured data (fig. 7.46a), including the width of the anti-bunching dip. More details on the fits can be found in the PhD thesis of M. Schwartz [124]. By subtracting dark counts from the measurement, the $g^{(2)}(0)$ reduces to 0.17 ± 0.04. This means that over 90 % of detected photons are on a single-photon level and stem from QD emission. This is very close to an expected $g^{(2)}(0)$ value of 0.293 (0.225 excluding dark counts) connected to the measured SNR under cw-resonant excitation (section 7.5.5).
For an application in quantum computation, the on-demand emission and detection of single photons is crucial. For the deterministic generation of high-quality photons with vanishing effects of dephasing, pulsed resonance fluorescence has been anticipated as the optimum [58]. Therefore, an on-chip auto-correlation measurement was also performed using pulsed resonant excitation fig. 7.46b. The excitation power was chosen slightly below the π-pulse. At zero time delay the peak is suppressed, showing a clear anti-bunching behavior. The $g^{(2)}(0)$ value is calculated by comparing the area of the peak at zero delay with the average area of all other peaks by adding up the intensity in each peak in a time interval of 2.56 ps around the maximum. The raw value of the calculated $g^{(2)}(0)$ is 0.59 ± 0.06 and 0.41 ± 0.04 after correction for detector dark counts. This proves that 77 % of detected photons stem from single-photon emission. The smaller $g^{(2)}(0)$ value in case of cw excitation is caused by the smaller line width (500 MHz compared to 15 GHz) and therefore more efficient excitation of the QD with less relative stray light [69]. Since the intensity of the QD emission is smaller for pulsed excitation, the relative DCR is higher than for CW excitation. [SSR+18]

This shows that the full on-chip HBT setup, which includes all three relevant components monolithically on the same chip, can be successfully operated on the single-photon level under resonant cw as well as on-demand under pulsed excitation. The SNR was successfully improved by the addition of an on-chip Al/AlN cover to the detectors and to the sides to the WG. This demonstration is the first successful monolithic on-chip integration of a full quantum-photonic circuit.

7.6 Summary

In this chapter, a fully integrated on-chip Hanbury-Brown Twiss setup was successfully demonstrated on a full monolithic GaAs-platform for the first time. To make this possible, the influence and reduction of noise counts was investigated. By using a fiber-coupled setup, the number of dark counts can be reduced by two orders of magnitude over a free-space setup. Thermal noise can be reduced by the filtering of thermal photons at the cold stage. Stray light was shown to have a position-dependent and a position-independent component. It was found that an on-detector Al/AlN cover is efficient in shielding the detector from the top, which reduces the amount of stray light by a factor of 2.6. However, it was found that a significant amount of stray light photons are guided in the substrate. To further reduce stray light, the substrate backside can be modified. A stray light suppression in the order of

2 to 6 can be achieved by using a dual-side instead of a single-side polished substrate. A small inclination of the excitation laser to guide stray light away from the detector showed no significant improvement compared to a perpendicular excitation of the QD. The emission and detection of above-band and resonantly-excited QD photons was demonstrate on-chip utilizing a straight WG. To improve the SNR, it was found that the total length of waveguides should be kept as short as possible if a WG with an attenuation $>3.5 \pm 0.2\,\mathrm{dB/mm}$ is used. For a lower attenuation, an optimal distance for QD excitation of $\approx 1100\,\mu\mathrm{m}$ to the detector was found. The fully on-chip HBT sample was built up on a single-side polished substrate. An on-chip beam splitter with a splitting ratio of $\frac{56.7}{43.3} \pm 4.4\%$ was implemented alongside SNSPDs with a detection efficiency of up to $47.5 \pm 15.7\,\%$. By the use of Al/AlN as covers on the detector and as large patches at either side of the WG, stray light is partly blocked from entering the substrate. An improvement of the SNR for a single QD under resonant excitation from 0.48 to 16.2 was achieved using large covers. This enabled the first demonstration of a full photonic chip including single-photon sources, WG-based photonic logic and single-photon detectors on a single chip operating on a single-photon level. The single-photon operation was demonstrated under cw as well as pulsed-resonant excitation in a free-space setup at a temperature of 4 K without the utilization of time gating techniques and spectral filtering. The monolithic approach of this thesis offers a method to straightforward realize large-scale quantum-photonic chips with a large range of capabilities by adding additional elements like phase shifters [60] and multiple sources of indistinguishable single-photons [62], which can be implemented on GaAs directly. By an implementation of all studied suppression techniques, the SNR and fidelity of the single-photon operation can be further increased. To further decrease stray light, the covers can be increased and put closer to the WG, in addition, to the use of a dual-side polished substrate.

8 Conclusions

In this thesis, SNSPDs for the monolithic integration of a fully integrated quantum-photonic circuit were investigated. GaAs was used as basic material system since it allows the integration of high-quality QD single-photon sources and the integration of a passive on-chip logic. The SNSPDs were made from thin NbN films to allow the operation at a temperature of 4 K. The growth of uniform NbN films on GaAs, which are required for efficient SNPSDs, is challenging due to a low thermal stability of the substrate, a significant lattice mismatch and interdiffusion of oxygen and arsenic from GaAs into the NbN. By the use of an AlN buffer layer and a careful optimization of deposition parameters a $T_C = 12.1$ K, $\Delta T_C = 1.1$ K and $RRR = 0.9$ was reached for a 5.5 nm thick NbN film. This is a significant improvement in comparison to previously reported T_C values for thin NbN films on GaAs (8.2 K [81], 8.7 K [80], 9.4 K [189]). The AlN buffer acts as an effective diffusion barrier and improves the NbN film quality at a thickness as small as 2.5 nm. The optimal AlN thickness was found to be 12 nm. In addition, the obtained $T_C > 10.5$ K for all investigated AlN thicknesses should be sufficient for a SNSPD operation at 4.2 K. This opens the scope for integrated SNSPDs on AlN-based photonic circuit using sputtered films. AlN films sputtered at room temperature have a $\varepsilon_r = 9.9 \pm 0.3$ and $\tan\delta < 0.005$ and were successfully used as a dielectric for the integration of a high pass filter with a 3 dB bandwidth of 5.6 GHz. This demonstrates that the sputtered AlN films have a sufficient quality to be used as a dielectric for on-chip RF circuits.

A double-spiral SNSPD design and a novel concept to integrate such a detector onto a single-mode waveguide using a tapered structure was introduced. The influence of an AlN buffer layer for NbN SNSPDs on GaAs was investigated by fabricating SNSPDs with and without buffer layer. The buffered SNSPD had a from 8.7 K to 9.3 K increased T_C and a significantly increased I_C from 8.8 µA to 16.5 µA in comparison to a SNSPD without buffer layer. This translates to a significantly increased detection efficiency of the buffered detector over the full investigated spectrum from 400 to 1700 nm in the order of one magnitude for an identical bias level. At a bias level of $I_B = 0.9 I_C$ at a wavelength of 900 nm (385 nm) a detection efficiency of 5 % (31 %) is reached with an AlN buffer layer. This is a significant improvement in comparison to the detection efficiency for a SNSPD directly on GaAs with 0.3 % (6 %). For the buffered detector the dark count rate was decreased by three orders of magnitude and the system timing jitter was reduced by 36 ps to a total of 72 ps. At $I_B = 0.95 I_C$ the buffered detector was found to be in the deterministic detection regime, which is characterized by an intrinsic detection efficiency close to unity, for wavelengths shorter than 796 nm, which is already close to the emission wavelength of 900 nm of the used InGaAs/GaAs QDs used for on-chip photonic experiments during this thesis. In summary, the AlN buffer layer significantly improves the detector properties. However, a further optimization is needed to broaden the deterministic detection regime towards 900 nm.

As a possibility to push the deterministic detection region towards smaller photon energies and consequently longer wavelengths, SNSPDs with variable thickness to minimize current crowding were proposed and tested in this work. Tight bends in the nanowire, as seen in the middle of a double-spiral SNSPD, cause a so-called current-crowding effect, which locally suppresses the critical current and prohibits an efficient biasing of the nanowire. In addition, the vortex entry barrier is suppressed in the bend, which increases the dark count rate. By significantly increasing the thickness of tight bends (11.3 nm) in comparison to the thickness of straight sections (4.3 nm) in a meander SNSPD, the critical current was increased by 30 ± 4 %. This increased the intrinsic detection efficiency to a maximum of 97 % at wavelength of 500 nm, which is an improvement by a factor of three in comparison to the witness device with uniform thickness with a maximum intrinsic detection efficiency of 32 %. The detection efficiency was increased for the full investigated spectral range from 500 to 1700 nm. A cutoff wavelength $\lambda_C = 1260$ nm was reached at $I_B = 0.90 I_C$, which is comparable to the λ_C obtainable for a NbN SNSPD in a single-spiral design [158]. In addition, variable thickness devices had a significantly decreased dark count rate in the order of one magnitude. This demonstrates that the proposed variable thickness design for SNSPDs is able to successfully reduce the current-crowding effect.

Many technologies in the field of quantum-photonic information technologies require the characterization of photon-number states. For linear optical quantum computing after the KLM scheme, detectors with a photon-number resolution of 4 photons are required. Standard SNSPDs only allow the detection of one photon at a time. To achieve photon-number resolution using SNSPDs, a concept proposed in [101] was used. In this concept, photon-number resolution in a SNSPD is obtained by a local shunting of the nanowire. Photon-number resolving detectors were designed and fabricated on sapphire as well as on GaAs substrate with two slightly different designs. The design of the SNSPD on GaAs is an evolution of the detector design on sapphire that allows waveguide integration. For the first time, 3- and 4-photon resolving detectors in a series design were demonstrated at temperatures ≥4.0 K. The timing jitter for both detectors was in the range of 130 ps. The measured statistics of the 3-pixel PNR-SNSPD on sapphire accurately represented the statistics of the used pulsed fs-laser with a repetition rate of 100 MHz and $\lambda = 1550$ nm, up to a mean photon number of 60 photons/pulse at a detection efficiency of 8.5 %. This is the first demonstration of a series PNR-SNSPD on sapphire. The intrinsic detection efficiency of the 4-pixel PNR-SNSPD on GaAs was estimated to 77 % at an $I_B = 0.9 I_C$. The statistics of a ps laser with a repetition rate of 76 MHz and $\lambda = 900$ nm was accurately represented up to a mean photon number of 5 photons/pulse at detection efficiency of 22.7 ± 3.0 %. The dynamic range and timing jitter was found to be limited by the readout and not intrinsically by the SNSPDs. A 4-photon number resolution was also found for a variety of light sources: a cw laser, a LED and a halogen light source and for all investigated wavelengths. A spectral range of 740 to 1550 nm was used to investigate the photon-number response of detectors. The response of the detector on GaAs to an almost ideal InGaAs/GaAs single-photon source was recorded on the one and two photon level and compared to a pulsed laser source with the same intensity. This demonstration showed the ability of the developed detector to directly discriminate a sub-Poissoninan from a Poissonian light source. This demonstration shows the practical benefit of a PNR-SNSPD for integrated quantum-photonic information systems. With the longer absorption length on a waveguide, together with an increase in the intrinsic detection efficiency using a variable thickness approach, a detection efficient close to unity is in reach for a waveguide integrated PNR-SNSPD with identical film quality.

Since previous demonstrations of the integration of QD with detectors on a single-chip [108] [72] severely suffered from stray light from the pump laser, the influence and suppression of noise counts for a SNSPD was investigated. It was found that the use of a fiber-coupled setup reduces the amount of dark counts by two orders of magnitude. Stray light of the excitation laser was found to be the main source of false detector counts and was found reaching the detector directly as well as through the substrate. The introduction of a Al/AlN cover layer directly on the detector was found to effectively shield the SNSPDs from direct stray light, which reduced the total amount or stray light on the detector by a factor of 2.6. By the use of a dual-side polished substrate instead of a single-side substrate, stray light was further reduced by a factor of 2 to 6. A small inclination of the pump laser to guide stray light away from the detector showed no significant improvement in comparison to a standard perpendicular orientation. By integrating a SNSPD onto a single-mode WG with integrated QDs, the emission, guiding and detection of above-band and resonantly excited QD photons was demonstrated. The SNSPD was implemented in a double-spiral design onto the broad side of a 250 µm long linear taper. It was found that the best signal-to-noise ratio is obtained by an excitation of the QD as close as possible to the start of the taper, as long as the waveguide attenuation is $>3.5 \pm 0.2$ dB/mm. For a lower waveguide attenuation, the optimal excitation distance was found to be $\approx$1100 µm.

A full Hanbury-Brown and Twiss setup was monolithically integrated into a single-GaAs chip using InGaAsGaAs QD single-photon sources, single-mode waveguides, a waveguide-based directional beam splitter and two SNSPDs. To further suppress stray light, planar Al/AlN covers were deterministically placed on SNSPDs and next to the waveguides. For the integrated SNSPDs a maximal detection efficiency of 47.5 ± 15.7 % was obtained. The waveguides were proven to be single-mode with an attenuation of 9.3 ± 1.1 dB/mm. With a splitting ratio of $56.7/43.3 \pm 4.4$% for the beam splitter at the QD emission wavelength (876.1 nm) a ratio close to 50/50 was achieved. The QD showed a fine structure splitting and was characterized off-chip to have an emission with a clear single-photon character with a $g^{(2)}(0) = 0.08 \pm 0.03$ for pulsed resonant excitation. In a TCSPC measurement using the on-chip SNSPDs, a distinct beating between the fine-structure components is visible. The used on-chip covers improved the on-chip signal-to-noise ratio from initially 0.48 to 16.2. The execution of a fully on-chip Hanbury-Brown and Twiss experiment proved the operation of all on-chip elements on the single-photon level. For cw resonant excitation of the QD, a $g^{(2)}(0) = 0.24 \pm 0.06$ was found for the raw measurement data. After subtraction of dark counts a $g^{(2)}(0) = 0.17 \pm 0.04$ is reached. This shows that over 90 % of detected photons on-chip stem from single-photon emission. Under pulsed resonant excitation a $g^{(2)}(0) = 0.59 \pm 0.06$ was found for the raw data and $g^{(2)}(0) = 0.43 \pm 0.05$ after subtraction of dark counts. The larger $g^{(2)}(0)$ for pulsed excitation can be explained by a lower excitation efficiency of the QD due to a larger line width of the excitation laser. Nevertheless, the dark count corrected data shows that over 75 % of detected photons stem from single-photon emission. This proves a successful on-demand operation of the on-chip setup. This demonstration is the first successful demonstration of a full quantum-photonic chip on a scalable monolithic platform. The shown experiment demonstrated the successful cooperative operation of all basic elements necessary to build up complex quantum-photonic logic on a single chip. This was achieved using a free space cryogenic setup at a temperature of 4 K without the utilization of time gating techniques or spectral filtering. The potential of the GaAs material system as a monolithic platform for quantum information technologies was demonstrated.

As the presented results are a significant break through for the GaAs photonic platform, consecutive steps that have to be taken for the integration of more complex circuits can be derived. An improvement of the on-chip $g^{(2)}(0)$ can be achieved by further reducing stray light and dark counts. Stray light can be further reduced by an extension of the Al/AlN cover over the full chip and by the use of an absorption layer or a Bragg reflector underneath the waveguide. Thermal noise can be reduced by the introduction of cold spectral filters. Dark counts can be reduced by a lower operation temperature or a fiber-based setup. The efficiencies of the QD and SNSPDs need to be improved to reach an accumulated efficiency above 2/3, which is required for linear-optical quantum-computing [226]. The QD coupling efficiency can be improved by an embedding of the QD locally into a photonic crystal waveguide. The efficiency of the waveguide SNSPDs can be improved by the use of a variable thickness design to increase the intrinsic detection efficiency. This further provides a larger freedom for the full detector design and allows for a further optimization of the absorption efficiency. As a next step, a fully on-chip Hong-Ou-Mandel experiment using two independent QD single-photon sources has to be demonstrated, to show the deterministic on-chip operation with indistinguishable photons which is necessary for the implementation of a photonic qubit. Furthermore, the developed 4-pixel photon-number resolving SNSPDs should be demonstrated on an integrated photonic chip.

Nomenclature

Acronyms

ABS	absorption efficiency
ADC	analog digital converter
AFM	atomic force microscopy
APD	avalanche photo diode
ASIC	application specific integrated circuit
BS	beam splitter
CR	count rate
cw	continuous wave
DCR	dark count rate
DE	detection efficiency
EBL	electron beam lithography
EMG	exponentially modified Gauss function
FSS	fine-structure splitting
HBT	Hanbury-Brown and Twiss
HOM	Hong-Ou-Mandel effect
HSQ	hydrogen silsesquioxane, a negative tone electron beam resist
IC	integrated circuit
ICP	inductively coupled plasma
IDE	intrinsic detection efficiency
IRF	instrumental response function
KLM	Knill, Laflamme and Milburn
LOQC	linear optical quantum computation

MMI	multimode interferometer
NA	numerical aperture
PL	photo luminescence
PM	power meter
PMMA	poly(methyl methacrylate), a positive tone electron beam resist
PN	photon number
PNR	photon-number resolution
PRF	pulse rate frequency
QD	quantum dot
QPIC	quantum-photonic integrated-circuit
RF	radio frequency
RIE	reactive ion etching
SDE	system detection efficiency
SEM	scanning electron microscope
SFWM	spontaneous four wave mixing
SNR	signal-to-noise ratio
SNSPD	superconducting nanowire single-photon detector
SOI	silicon on insulator
SPDC	spontaneous down conversion
TCSPC	time-correlated single-photon counting
TE	transverse electric
TES	transition edge sensor
TM	transverse magnetic
WG	waveguide
WG-SNSPD	waveguide-integrated superconducting nanowire single-photon detector

Constants

c	speed of light in vacuum: 299 792 458 m/s
h	Planck's constant: $6.626\,070\,15 \times 10^{-34}$ J/s
k_B	Boltzmann constant: $1.380\,648\,52 \times 10^{-23}$ J/K
ε_0	vacuum permittivity: 8.854 187 817 A s m/V
π	Ludolph's constant: 3.141 59...

Latin Symbols

D	electron diffusion coefficient
$\mathrm{DE}_{\mathrm{plateau}}$	detection efficiency in the deterministic detection regime
d	thickness
d_{laser}	spot diameter of the laser on the sample
E	electric field
E_{min}	minimal photon energy for deterministic single-photon detection
E_{therm}	thermal energy
E_G	energy gap
f	frequency
$f(x)$	function of x
f_0	resonance frequency
$g^{(2)}$	second order correlation function
$g^{(n)}$	Glauber function to the n-th order
I	intensity
I_{C}	critical current
$I_{\mathrm{C}}^{\mathbf{GL}}(0)$	critical current for $T \mapsto 0$ K calculated for the Ginzburg-Landau temperature dependence
I_{dep}	depairing current
I_{lk}	leakage current
I_{sp}	sputter current

I_r	re-trapping current
j_C	critical current density
j_r	re-trapping current density
k	wavenumber
k_{asym}	propagation constant of the assymmetric mode
k_{sym}	propagation constant of the symmetric mode
$KL(T)$	Kupriyanov-Lukichev correction factor
L	interaction length or inductance
l	length
ℓ_S	electron mean free-path
$L_\square$	sheet inductance
L_{geo}	geometric inductance
L_{kin}	kinetic inductance
m_e^*	effective electron mass
n	arbitrary natural number or refractive index
N	defined number
$n(r)$	refractive index profile
N_x	defined number of x
N_0	electron density of states
p	fit parameter to describe a power law dependence
p_{Ar}	argon partial pressure
p_{N_2}	nitrogen partial pressure
$p_{\textbf{total}}$	total system pressure
r	radius
R	reflected power or reduction factor
$R(T)$	temperature-dependet resistance
$R_\square$	sheet resistance
R_{300}	room temperature resistance
R_n	residual resistance

R_P	parallel resistor (shunt resistor)
R_S	series resistor
RRR	residual resistance ratio
SR	splitting ratio
T	absolute temperature or transmitted power
t	time
T_C	critical temperature
T_B	bath temperature
T_m	melting temperature
w	width
X	exciton
X^-	negatively charged trion
X^+	positively charged trion
XX	biexciton
Z	impedance

Other Symbols

α	excitation angle
γ	device- and material-dependent proportionality factor
Γ_1	radiative decay rate
Γ_2	damping rate
ΔT_C	width of superconducting transition
Δ	superconducting energy gap
$\tan\delta$	dielectric loss factor
ε_r	relative permittivity
η	detection efficiency as fit parameter
η_{count}	probability that an excited QD leads to the detection of a photon
κ	attenuation coefficient (complex part of the refractive index)

Λ	pearl length
λ	wavelength
λ_C	cutoff wavelength
λ_L	London penetration depth
μ	average number of photons per pulse
ξ_{GL}	Ginzburg-Landau coherence length
$\sqcap(x)$	rectangular function
ρ_{300}	room temperature resistivity
ρ_{defects}	resistivity caused by collisions with defects in the crystal lattice
ρ_{electron}	resistivity caused by collisions with electrons
ρ_n	residual resistivity
ρ_{phonon}	resistivity caused by collisions with phonons
$\rho_{\text{tunneling}}$	tunneling resistivity
σ	standard deviation
ς	efficiency of the quasiparticle multiplication process
τ	characteristic time constant
τ_{th}	thermalization time constant
ϕ	intitial phase
ρ	resistivity
Ω	rabi frequency

Gates

CNOT	controlled not gate
H	Hadamard gate
NAND	NOT-AND gate
T	phase shift gate

Chemical Compositions

AlGaAs	aluminum gallium arsenide
Ar^+	argon ion
Ar	argon
As	arsenide
GaAs	gallium arsenide
InAs	indium gallium arsenide
InGaAs	indium gallium arsenide
$LiNbO_3$	lithium niobate
MoSi	molybdenum silicide
N_2	nitrogen
Nb	niobium
NbN	niobium nitride
NbTiN	niobium titanium nitride
O_2	oxygen
SF_6	sulfur hexafluoride
$SiCl_4$	silicon tetrachloride
SiH	silicon monohydride
SiO	silicon oxide
TaN	tantalum nitride
Ti	titianium
WSi	tungsten silicide

List of Figures

List of Tables

Bibliography

[1] Max Planck. Über das Gesetz der Energieverteilung im Normalspektrum. In Hans-Georg Schöpf, editor, *Von Kirchhoff bis Planck*, pages 178–191. Vieweg+Teubner Verlag, Wiesbaden, 1978.

[2] Lov K. Grover. A fast quantum mechanical algorithm for database search. In Gary L. Miller, editor, *Proceedings of the twenty-eighth annual ACM symposium on Theory of computing - STOC '96*, pages 212–219, New York, New York, USA, 1996. ACM Press.

[3] P. W. Shor. Algorithms for quantum computation: discrete logarithms and factoring. In *Proceedings 35th Annual Symposium on Foundations of Computer Science*, pages 124–134. IEEE Comput. Soc. Press, 20-22 Nov. 1994.

[4] Richard P. Feynman. Simulating physics with computers. *International Journal of Theoretical Physics*, 21(6-7):467–488, 1982.

[5] D. Deutsch. Quantum Theory, the Church-Turing Principle and the Universal Quantum Computer. *Proceedings of the Royal Society A: Mathematical, Physical and Engineering Sciences*, 400(1818):97–117, 1985.

[6] Sergio Boixo, Sergei V. Isakov, Vadim N. Smelyanskiy, Ryan Babbush, Nan Ding, Zhang Jiang, Michael J. Bremner, John M. Martinis, and Hartmut Neven. Characterizing quantum supremacy in near-term devices. *Nature Physics*, 14(6):595–600, 2018.

[7] Ekert. Quantum cryptography based on Bell's theorem. *Physical Review Letters*, 67(6):661–663, 1991.

[8] Nicolas Gisin, Grégoire Ribordy, Wolfgang Tittel, and Hugo Zbinden. Quantum cryptography. *Rev. Mod. Phys.*, 74(1):145–195, 2002.

[9] Charles H. Bennett and David P. DiVincenzo. Quantum information and computation. *Nature*, 404(6775):247, 2000.

[10] Thaddeus D. Ladd, Fedor Jelezko, Raymond Laflamme, Yasunobu Nakamura, Christopher Monroe, and Jeremy Lloyd O'Brien. Quantum computers. *Nature*, 464(7285):45, 2010.

[11] Chao Song, Kai Xu, Wuxin Liu, Chui-Ping Yang, Shi-Biao Zheng, Hui Deng, Qiwei Xie, Keqiang Huang, Qiujiang Guo, Libo Zhang, Pengfei Zhang, Da Xu, Dongning Zheng, Xiaobo Zhu, H. Wang, Y-A Chen, C-Y Lu, Siyuan Han, and Jian-Wei Pan. 10-Qubit Entanglement and Parallel Logic Operations with a Superconducting Circuit. *Physical Review Letters*, 119(18):180511, 2017.

[12] Pieter Kok, W. J. Munro, Kae Nemoto, T. C. Ralph, Jonathan P. Dowling, and G. J. Milburn. Linear optical quantum computing with photonic qubits. *Rev. Mod. Phys.*, 79(1):135–174, 2007.

[13] E. Knill, R. Laflamme, and G. J. Milburn. A scheme for efficient quantum computation with linear optics. *Nature*, 409(6816):46, 2001.

[14] Justin B. Spring, Benjamin J. Metcalf, Peter C. Humphreys, W. Steven Kolthammer, Xian-Min Jin, Marco Barbieri, Animesh Datta, Nicholas Thomas-Peter, Nathan K. Langford, Dmytro Kundys, James C. Gates, Brian J. Smith, Peter G. R. Smith, and Ian A. Walmsley. Boson sampling on a photonic chip. *Science*, 339(6121):798–801, 2013.

[15] Scott Aaronson and Alex Arkhipov. The computational complexity of linear optics. In Lance Fortnow and Salil Vadhan, editors, *Proceedings of the 43rd ACM Symposium on Theory of Computing*, page 333, New York, NY, 2011. ACM.

[16] Peter Michler, editor. *Quantum Dots for Quantum Information Technologies*. Nano-Optics and Nanophotonics. Springer International Publishing, Cham, 2017.

[17] J. L. O'Brien, G. J. Pryde, A. G. White, T. C. Ralph, and D. Branning. Demonstration of an all-optical quantum controlled-NOT gate. *Nature*, 426(6964):264–267, 2003.

[18] Sara Gasparoni, Jian-Wei Pan, Philip Walther, Terry Rudolph, and Anton Zeilinger. Realization of a photonic controlled-NOT gate sufficient for quantum computation. *Physical Review Letters*, 93(2):020504, 2004.

[19] Jonathan P. Dowling. Quantum optical metrology – the lowdown on high-N00N states. *Contemporary Physics*, 49(2):125–143, 2008.

[20] M. Müller, H. Vural, C. Schneider, A. Rastelli, O. G. Schmidt, S. Höfling, and P. Michler. Quantum-Dot Single-Photon Sources for Entanglement Enhanced Interferometry. *Phys. Rev. Lett.*, 118(25):257402, 2017.

[21] Alán Aspuru-Guzik and Philip Walther. Photonic quantum simulators. *Nature Physics*, 8:285–, 2012.

[22] Jack S. Kilby. Miniaturized electronic circuits, 1964. US3138743A.

[23] Robert N. Noyce. Semiconductor device-and-lead structure, 1961. US2981877A.

[24] P. O. Boykin, T. Mor, M. Pulver, V. Roychowdhury, and F. Vatan. On universal and fault-tolerant quantum computing: a novel basis and a new constructive proof of universality for Shor's basis. In *40th Annual Symposium on Foundations of Computer Science (Cat. No.99CB37039)*, pages 486–494. IEEE Comput. Soc, 17-19 Oct. 1999.

[25] Hong, Ou, and Mandel. Measurement of subpicosecond time intervals between two photons by interference. *Physical Review Letters*, 59(18):2044–2046, 1987.

[26] Alberto Politi, Martin J. Cryan, John G. Rarity, Siyuan Yu, and Jeremy L. O'Brien. Silica-on-Silicon Waveguide Quantum Circuits. *Science*, 320(5876):646–649, 2008.

[27] Alberto Politi, Jonathan C. F. Matthews, and Jeremy L. O'Brien. Shor's quantum factoring algorithm on a photonic chip. *Science*, 325(5945):1221, 2009.

[28] Joshua W. Silverstone, Damien Bonneau, Jeremy L. O'Brien, and Mark G. Thompson. Silicon Quantum Photonics. *IEEE Journal of Selected Topics in Quantum Electronics*, 22(6):390–402, 2016.

[29] Jacques Carolan, Christopher Harrold, Chris Sparrow, Jeremy L. Mart\textquoterightBrien, and Anthony Laing. Universal linear optics. *Science*, 349(6249):711–716, 2015.

[30] M. Poot and H. X. Tang. Characterization of optical quantum circuits using resonant phase shifts. *Applied Physics Letters*, 109(13):131106, 2016.

[31] Nicholas J. D. Martinez, Michael Gehl, Christopher T. Derose, Andrew L. Starbuck, Andrew T. Pomerene, Anthony L. Lentine, Douglas C. Trotter, and Paul S. Davids. Single photon detection in a waveguide-coupled Ge-on-Si lateral avalanche photodiode. *Optics express*, 25(14):16130–16139, 2017.

[32] Brice Calkins, Paolo L. Mennea, Adriana E. Lita, Benjamin J. Metcalf, W. Steven Kolthammer, Antia Lamas-Linares, Justin B. Spring, Peter C. Humphreys, Richard P. Mirin, James C. Gates, Peter G. R. Smith, Ian A. Walmsley, Thomas Gerrits, and Sae Woo Nam. High quantum-efficiency photon-number-resolving detector for photonic on-chip information processing. *Optics express*, 21(19):22657–22670, 2013.

[33] J. P. Sprengers, A. Gaggero, D. Sahin, S. Jahanmirinejad, G. Frucci, F. Mattioli, R. Leoni, Johannes Beetz, M. Lermer, M. Kamp, S. Höfling, R. Sanjines, and A. Fiore. Waveguide superconducting single-photon detectors for integrated quantum photonic circuits. *Applied Physics Letters*, 99(18):181110, 2011.

[34] Tsung-Ju Lu, Michael Fanto, Hyeongrak Choi, Paul Thomas, Jeffrey Steidle, Sara Mouradian, Wei Kong, Di Zhu, Hyowon Moon, Karl Berggren, Jeehwan Kim, Mohammad Soltani, Stefan Preble, and Dirk Englund. Aluminum nitride integrated photonics platform for the ultraviolet to visible spectrum. *Optics express*, 26(9):11147–11160, 2018.

[35] Giulia Enrica Digeronimo, Maurangelo Petruzzella, Simone Birindelli, Rosalinda Gaudio, Sartoon Fattah Poor, Frank W. M. van Otten, and Andrea Fiore. Integration of single-photon sources and detectors on GaAs. In *Photonics*, volume 3, page 55, 2016.

[36] Simone Ferrari, Carsten Schuck, and Wolfram Pernice. Waveguide-integrated superconducting nanowire single-photon detectors. *Nanophotonics*, 7(11):1725–1758, 2018.

[37] F. Marsili, V. B. Verma, J. A. Stern, S. Harrington, A. E. Lita, T. Gerrits, I. Vayshenker, B. Baek, M. D. Shaw, R. P. Mirin, and S. W. Nam. Detecting single infrared photons with 93% system efficiency. *Nature Photonics*, 7(3):210, 2013.

[38] Adriana E. Lita, Aaron J. Miller, and Sae Woo Nam. Counting near-infrared single-photons with 95% efficiency. *Optics Express*, 16(5):3032, 2008.

[39] Alessandro Restelli, Joshua C. Bienfang, and Alan L. Migdall. Single-photon detection efficiency up to 50% at 1310 nm with an InGaAs/InP avalanche diode gated at 1.25 GHz. *Applied Physics Letters*, 102(14):141104, 2013.

[40] Francesco Mattioli, Zili Zhou, Alessandro Gaggero, Rosalinda Gaudio, Saeedeh Jahanmirinejad, Döndü Sahin, Francesco Marsili, Roberto Leoni, and Andrea Fiore. Photon-number-resolving superconducting nanowire detectors. *Superconductor Science and Technology*, 28(10):104001, 2015.

[41] Robert H. Hadfield. Single-photon detectors for optical quantum information applications. *Nature Photonics*, 3(12):696–705, 2009.

[42] B. A. Korzh, Q-Y Zhao, S. Frasca, J. P. Allmaras, T. M. Autry, E. A. Bersin, M. Colangelo, G. M. Crouch, A. E. Dane, T. Gerrits, F. Marsili, G. Moody, E. Ramirez, J. D. Rezac, M. J. Stevens, E. E. Wollman, D. Zhu, P. D. Hale, K. L. Silverman, R. P. Mirin, S. W. Nam, M. D. Shaw, and K. K. Berggren. Demonstrating sub-3 ps temporal resolution in a superconducting nanowire single-photon detector.

[43] S. Bogdanov, M. Y. Shalaginov, A. Boltasseva, and V. M. Shalaev. Material platforms for integrated quantum photonics. *Optical Materials Express*, 7(1):111, 2017.

[44] Jianwei Wang, Stefano Paesani, Yunhong Ding, Raffaele Santagati, Paul Skrzypczyk, Alexia Salavrakos, Jordi Tura, Remigiusz Augusiak, Jeremy L. Man\textquoterightBrien, Anthony Laing, and Mark G. Thompson. Multidimensional quantum entanglement with large-scale integrated optics. *Science*, 2018.

[45] H. Jin, F. M. Liu, P. Xu, J. L. Xia, M. L. Zhong, Y. Yuan, J. W. Zhou, Y. X. Gong, W. Wang, and S. N. Zhu. On-chip generation and manipulation of entangled photons based on reconfigurable lithium-niobate waveguide circuits. *Physical Review Letters*, 113(10):103601, 2014.

[46] Jianwei Wang, Damien Bonneau, Matteo Villa, Joshua W. Silverstone, Raffaele Santagati, Shigehito Miki, Taro Yamashita, Mikio Fujiwara, Masahide Sasaki, Hirotaka Terai, Michael G. Tanner, Chandra M. Natarajan, Robert H. Hadfield, Jeremy L. O'Brien, and Mark G. Thompson. Chip-to-chip quantum photonic interconnect by path-polarization interconversion. *Optica*, 3(4):407, 2016.

[47] Sven Ramelow, Alessandro Farsi, Stéphane Clemmen, Daniel Orquiza, Kevin Luke, Michal Lipson, and Alexander L. Gaeta. Silicon-Nitride Platform for Narrowband Entangled Photon Generation.

[48] Justin B. Spring, Patrick S. Salter, Benjamin J. Metcalf, Peter C. Humphreys, Merritt Moore, Nicholas Thomas-Peter, Marco Barbieri, Xian-Min Jin, Nathan K. Langford, W. Steven Kolthammer, Martin J. Booth, and Ian A. Walmsley. On-chip low loss heralded source of pure single photons. *Optics express*, 21(11):13522–13532, 2013.

[49] D. N. Klyshko, A. N. Penin, and B. F. Plkovnikov. Parametric Luminescence and Light Scattering by Polaritons. *Journal of Experimental and Theoretical Physics Letters*, 11:5, 1970.

[50] Marco Bazzan and Cinzia Sada. Optical waveguides in lithium niobate: Recent developments and applications. *Applied Physics Reviews*, 2(4):040603, 2015.

[51] Jan Philipp Höpker, Moritz Bartnick, Evan Meyer-Scott, Frederik Thiele, Torsten Meier, Tim Bartley, Stephan Krapick, Nicola M. Montaut, Matteo Santandrea, Harold Herrmann, Sebastian Lengeling, Raimund Ricken, Viktor Quiring, Adriana E. Lita, Varun B. Verma, Thomas Gerrits, Sae Woo Nam, and Christine Silberhorn. Towards integrated superconducting detectors on lithium niobate waveguides. In Mario Agio, Kartik Srinivasan, and Cesare Soci, editors, *Quantum Photonic Devices*, page 7. SPIE, 06.08.2017 - 10.08.2017.

[52] Daniel White, Artur Branny, Robert J. Chapman, Raphaël Picard, Mauro Brotons-Gisbert, Andreas Boes, Alberto Peruzzo, Cristian Bonato, and Brian D. Gerardot.

Atomically-thin quantum dots integrated with lithium niobate photonic chips [Invited]. *Optical Materials Express*, 9(2):441, 2019.

[53] P. Michler, A. Kiraz, C. Becher, W. V. Schoenfeld, P. M. Petroff, Lidong Zhang, E. Hu, and A. Imamoglu. A Quantum Dot Single-Photon Turnstile Device. *Science*, 290(5500):2282–2285, 2000.

[54] L. J. Rogers, K. D. Jahnke, T. Teraji, L. Marseglia, C. Müller, B. Naydenov, H. Schauffert, C. Kranz, J. Isoya, L. P. McGuinness, and F. Jelezko. Multiple intrinsically identical single-photon emitters in the solid state. *Nature communications*, 5:4739, 2014.

[55] A. Sipahigil, R. E. Evans, D. D. Sukachev, M. J. Burek, J. Borregaard, M. K. Bhaskar, C. T. Nguyen, J. L. Pacheco, H. A. Atikian, C. Meuwly, R. M. Camacho, F. Jelezko, E. Bielejec, H. Park, M. Lončar, and M. D. Lukin. An integrated diamond nanophotonics platform for quantum-optical networks. *Science*, 354(6314):847–850, 2016.

[56] Tim Schröder, Sara L. Mouradian, Jiabao Zheng, Matthew E. Trusheim, Michael Walsh, Edward H. Chen, Luozhou Li, Igal Bayn, and Dirk Englund. Quantum nanophotonics in diamond [Invited]. *Journal of the Optical Society of America B*, 33(4):B65, 2016.

[57] M. Arcari, I. Söllner, A. Javadi, S. Lindskov Hansen, S. Mahmoodian, J. Liu, H. Thyrrestrup, E. H. Lee, J. D. Song, S. Stobbe, and P. Lodahl. Near-Unity Coupling Efficiency of a Quantum Emitter to a Photonic Crystal Waveguide. *Phys. Rev. Lett.*, 113(9):093603, 2014.

[58] Yu-Ming He, Yu He, Yu-Jia Wei, Dian Wu, Mete Atatüre, Christian Schneider, Sven Höfling, Martin Kamp, Chao-Yang Lu, and Jian-Wei Pan. On-demand semiconductor single-photon source with near-unity indistinguishability. *Nature nanotechnology*, 8(3):213–217, 2013.

[59] H. Inoue, K. Hiruma, K. Ishida, T. Asai, and H. Matsumura. Low loss GaAs optical waveguides. *IEEE Transactions on Electron Devices*, 32(12):2662–2668, 1985.

[60] Christof P. Dietrich, Andrea Fiore, Mark G. Thompson, Martin Kamp, and Sven Höfling. GaAs integrated quantum photonics: Towards compact and multi–functional quantum photonic integrated circuits. *Laser & Photonics Reviews*, 10(6):870–894, 2016.

[61] Leonardo Midolo, Sofie L. Hansen, Weili Zhang, Camille Papon, Rüdiger Schott, Arne Ludwig, Andreas D. Wieck, Peter Lodahl, and Søren Stobbe. Electro-optic routing of photons from a single quantum dot in photonic integrated circuits. *Optics Express*, 25(26):33514, 2017.

[62] Raj B. Patel, Anthony J. Bennett, Ian Farrer, Christine A. Nicoll, David A. Ritchie, and Andrew J. Shields. Two-photon interference of the emission from electrically tunable remote quantum dots. *Nature Photonics*, 4:632–, 2010.

[63] Javier Martín-Sánchez, Rinaldo Trotta, Antonio Mariscal, Rosalía Serna, Giovanni Piredda, Sandra Stroj, Johannes Edlinger, Christian Schimpf, Johannes Aberl, Thomas Lettner, Johannes Wildmann, Huiying Huang, Xueyong Yuan, Dorian Ziss,

Julian Stangl, and Armando Rastelli. Strain-tuning of the optical properties of semiconductor nanomaterials by integration onto piezoelectric actuators. *Semiconductor Science and Technology*, 33(1):013001, 2018.

[64] K. Konthasinghe, M. Peiris, Y. Yu, M. F. Li, J. F. He, L. J. Wang, H. Q. Ni, Z. C. Niu, C. K. Shih, and A. Muller. Field-Field and Photon-Photon Correlations of Light Scattered by Two Remote Two-Level InAs Quantum Dots on the Same Substrate. *Physical Review Letters*, 109(26), 2012.

[65] N. Prtljaga, R. J. Coles, J. O'Hara, B. Royall, E. Clarke, A. M. Fox, and M. S. Skolnick. Monolithic integration of a quantum emitter with a compact on-chip beamsplitter. *Applied Physics Letters*, 104(23):231107, 2014.

[66] Luca Sapienza, Henri Thyrrestrup, Søren Stobbe, Pedro David Garcia, Stephan Smolka, and Peter Lodahl. Cavity Quantum Electrodynamics with Anderson-Localized Modes. *Science*, 327(5971):1352–1355, 2010.

[67] U. Rengstl, M. Schwartz, T. Herzog, F. Hargart, M. Paul, S. L. Portalupi, M. Jetter, and P. Michler. On-chip beamsplitter operation on single photons from quasi-resonantly excited quantum dots embedded in GaAs rib waveguides. *Applied Physics Letters*, 107(2):021101, 2015.

[68] Maxim N. Makhonin, James E. Dixon, Rikki J. Coles, Ben Royall, Isaac J. Luxmoore, Edmund Clarke, Maxime Hugues, Maurice S. Skolnick, and A. Mark Fox. Waveguide Coupled Resonance Fluorescence from On-Chip Quantum Emitter. *Nano Letters*, 14(12):6997–7002, 2014.

[69] Mario Schwartz, Ulrich Rengstl, Thomas Herzog, Matthias Paul, Jan Kettler, Simone Luca Portalupi, Michael Jetter, and Peter Michler. Generation, guiding and splitting of triggered single photons from a resonantly excited quantum dot in a photonic circuit. *Opt. Express*, 24(3):3089–3094, 2016.

[70] Peter Schnauber, Johannes Schall, Samir Bounouar, Theresa Höhne, Suk-In Park, Geun-Hwan Ryu, Tobias Heindel, Sven Burger, Jin-Dong Song, Sven Rodt, and Stephan Reitzenstein. Deterministic Integration of Quantum Dots into on-Chip Multimode Interference Beamsplitters Using in Situ Electron Beam Lithography. *Nano Letters*, 18(4):2336–2342, 2018.

[71] A. Gaggero, S. Jahanmiri Nejad, F. Marsili, F. Mattioli, R. Leoni, D. Bitauld, D. Sahin, G. J. Hamhuis, R. Nötzel, R. Sanjines, and A. Fiore. Nanowire superconducting single-photon detectors on GaAs for integrated quantum photonic applications. *Applied Physics Letters*, 97(15):151108, 2010.

[72] G. Reithmaier, S. Lichtmannecker, T. Reichert, P. Hasch, K. Müller, M. Bichler, R. Gross, and J. J. Finley. On-chip time resolved detection of quantum dot emission using integrated superconducting single photon detectors. *Scientific Reports*, 3:1901, 2013.

[73] E. Murray, D. J. P. Ellis, T. Meany, F. F. Floether, J. P. Lee, J. P. Griffiths, G. A. C. Jones, I. Farrer, D. A. Ritchie, A. J. Bennett, and A. J. Shields. Quantum photonics hybrid integration platform. *Applied Physics Letters*, 107(17):171108, 2015.

[74] Iman Esmaeil Zadeh, Ali W. Elshaari, Klaus D. Jöns, Andreas Fognini, Dan Dalacu, Philip J. Poole, Michael E. Reimer, and Val Zwiller. Deterministic Integration of

Single Photon Sources in Silicon Based Photonic Circuits. *Nano Letters*, 16(4):2289–2294, 2016.

[75] Marcelo Davanco, Jin Liu, Luca Sapienza, Chen-Zhao Zhang, José Vin\'\icius Miranda Cardoso, Varun Verma, Richard Mirin, Sae Woo Nam, Liu Liu, and Kartik Srinivasan. Heterogeneous integration for on-chip quantum photonic circuits with single quantum dot devices. *Nature communications*, 8(1):889, 2017.

[76] D. J. P. Ellis, A. J. Bennett, C. Dangel, J. P. Lee, J. P. Griffiths, T. A. Mitchell, T.-K. Paraiso, P. Spencer, D. A. Ritchie, and A. J. Shields. Independent indistinguishable quantum light sources on a reconfigurable photonic integrated circuit. *Applied Physics Letters*, 112(21):211104, 2018.

[77] Alex D. Semenov, Gregory N. Gol'tsman, and Alexander A. Korneev. Quantum detection by current carrying superconducting film. *Physica C: Superconductivity and its Applications*, 351(4):349–356, 2001.

[78] G. N. Gol'tsman, O. Okunev, G. Chulkova, A. Lipatov, A. Semenov, K. Smirnov, B. Voronov, A. Dzardanov, C. Williams, and Roman Sobolewski. Picosecond superconducting single-photon optical detector. *Applied Physics Letters*, 79(6):705–707, 2001.

[79] Philipp Haas, Fabien Tran, and Peter Blaha. Calculation of the lattice constant of solids with semilocal functionals. *Physical Review B*, 79(8):1065, 2009.

[80] G. Reithmaier, J. Senf, S. Lichtmannecker, T. Reichert, F. Flassig, A. Voss, R. Gross, and J. J. Finley. Optimisation of NbN thin films on GaAs substrates for in-situ single photon detection in structured photonic devices. *Journal of Applied Physics*, 113(14):143507, 2013.

[81] Francesco Marsili, Alessandro Gaggero, Lianhe H. Li, Alessandro Surrente, Roberto Leoni, Francis Lévy, and Andrea Fiore. High quality superconducting NbN thin films on GaAs. *Superconductor Science and Technology*, 22(9):095013, 2009.

[82] A. Guillén-Cervantes, Z. Rivera-Alvarez, M. López-López, E. López-Luna, and I. Hernández-Calderón. GaAs surface oxide desorption by annealing in ultra high vacuum. *Thin Solid Films*, 373(1):159–163, 2000.

[83] Leon N. Cooper. Superconductivity in the Neighborhood of Metallic Contacts. *Phys. Rev. Lett.*, 6(12):689–690, 1961.

[84] Ya. V. Fominov and M. V. Feigel'man. Superconductive properties of thin dirty superconductor–normal-metal bilayers. *Physical Review B*, 63(9):1175, 2001.

[85] R. Schneider, B. Freitag, D. Gerthsen, K. S. Ilin, and M. Siegel. Structural, microchemical and superconducting properties of ultrathin NbN films on silicon. *Crystal Research and Technology*, 44(10):1115–1121, 2009.

[86] Jeremy L. O'Brien, Akira Furusawa, and Jelena Vučković. Photonic quantum technologies. *Nature Photonics*, 3(12):687, 2009.

[87] M. Schmidt, M. von Helversen, M. López, F. Gericke, E. Schlottmann, T. Heindel, S. Kück, S. Reitzenstein, and J. Beyer. Photon-Number-Resolving Transition-Edge Sensors for the Metrology of Quantum Light Sources. *Journal of Low Temperature Physics*, 2018.

[88] A. Meda, E. Losero, N. Samantaray, F. Scafirimuto, S. Pradyumna, A. Avella, I. Ruo-Berchera, and M. Genovese. Photon-number correlation for quantum enhanced imaging and sensing. *Journal of Optics*, 19(9):094002, 2017.

[89] F. E. Becerra, J. Fan, and A. Migdall. Photon number resolution enables quantum receiver for realistic coherent optical communications. *Nature Photonics*, 9(1):48, 2015.

[90] Roy J. Glauber. Coherent and Incoherent States of the Radiation Field. *Physical Review*, 131(6):2766–2788, 1963.

[91] Tim Thomay, Sergey V. Polyakov, Olivier Gazzano, Elizabeth Goldschmidt, Zachary D. Eldredge, Tobias Huber, Vivien Loo, and Glenn S. Solomon. Simultaneous, Full Characterization of a Single-Photon State. *Physical Review X*, 7(4), 2017.

[92] Clinton Cahall, Kathryn L. Nicolich, Nurul T. Islam, Gregory P. Lafyatis, Aaron J. Miller, Daniel J. Gauthier, and Jungsang Kim. Multi-photon detection using a conventional superconducting nanowire single-photon detector. *Optica*, 4(12):1534, 2017.

[93] Kathryn L. Nicolich, Clinton Cahall, Nurul T. Islam, Gregory P. Lafyatis, Jungsang Kim, Aaron J. Miller, and Daniel J. Gauthier. Universal Model for the Turn-on Dynamics of Superconducting Nanowire Single-Photon Detectors.

[94] Eric A. Dauler, Andrew J. Kerman, Bryan S. Robinson, Joel K.W. Yang, Boris Voronov, Gregory Goltsman, Scott A. Hamilton, and Karl K. Berggren. Photon-number-resolution with sub-30-ps timing using multi-element superconducting nanowire single photon detectors. *Journal of Modern Optics*, 56(2-3):364–373, 2009.

[95] Döndü Sahin, Alessandro Gaggero, Thang Ba Hoang, Giulia Frucci, Francesco Mattioli, Roberto Leoni, Johannes Beetz, Matthias Lermer, Martin Kamp, Sven Höfling, and Andrea Fiore. Integrated autocorrelator based on superconducting nanowires. *Optics express*, 21(9):11162–11170, 2013.

[96] Steffen Doerner, Artem Kuzmin, Stefan Wuensch, Ilya Charaev, and Michael Siegel. Operation of Multi-Pixel Radio-Frequency Superconducting Nanowire Single-Photon Detector Arrays. *IEEE Transactions on Applied Superconductivity*, page 1, 2016.

[97] S. Doerner, A. Kuzmin, S. Wuensch, I. Charaev, F. Boes, T. Zwick, and M. Siegel. Frequency-multiplexed bias and readout of a 16-pixel superconducting nanowire single-photon detector array. *Applied Physics Letters*, 111(3):032603, 2017.

[98] M. Hofherr, M. Arndt, K. Il'in, D. Henrich, M. Siegel, J. Toussaint, T. May, and H. Meyer. Time-Tagged Multiplexing of Serially Biased Superconducting Nanowire Single-Photon Detectors. *IEEE Transactions on Applied Superconductivity*, 23(3):2501205, 2013.

[99] Di Zhu, Qing-Yuan Zhao, Hyeongrak Choi, Tsung-Ju Lu, Andrew E. Dane, Dirk Englund, and Karl K. Berggren. A scalable multi-photon coincidence detector based on superconducting nanowires. *Nature nanotechnology*, 13(7):596–601, 2018.

[100] Aleksander Divochiy, Francesco Marsili, David Bitauld, Alessandro Gaggero, Roberto Leoni, Francesco Mattioli, Alexander Korneev, Vitaliy Seleznev, Nataliya Kaurova, Olga Minaeva, Gregory Gol'tsman, Konstantinos G. Lagoudakis, Moushab

Benkhaoul, Francis Lévy, and Andrea Fiore. Superconducting nanowire photon-number-resolving detector at telecommunication wavelengths. *Nature Photonics*, 2(5):302, 2008.

[101] Saeedeh Jahanmirinejad and Andrea Fiore. Proposal for a superconducting photon number resolving detector with large dynamic range. *Optics express*, 20(5):5017–5028, 2012.

[102] S. Jahanmirinejad, G. Frucci, F. Mattioli, D. Sahin, A. Gaggero, R. Leoni, and A. Fiore. Photon-number resolving detector based on a series array of superconducting nanowires. *Applied Physics Letters*, 101(7):072602, 2012.

[103] F. Mattioli, S. Jahanmirinejad, Z. Zhou, A. Gaggero, G. Frucci, D. Sahin, R. Leoni, and A. Fiore. Superconducting nanowires connected in series for photon number resolving functionality. *Journal of Physics: Conference Series*, 507(4):042024, 2014.

[104] Zili Zhou, Saeedeh Jahanmirinejad, Francesco Mattioli, Döndü Sahin, Giulia Frucci, Alessandro Gaggero, Roberto Leoni, and Andrea Fiore. Superconducting series nanowire detector counting up to twelve photons. *Optics express*, 22(3):3475–3489, 2014.

[105] D. Sahin, A. Gaggero, Z. Zhou, S. Jahanmirinejad, F. Mattioli, R. Leoni, J. Beetz, M. Lermer, M. Kamp, S. Höfling, and A. Fiore. Waveguide photon-number-resolving detectors for quantum photonic integrated circuits. *Applied Physics Letters*, 103(11):111116, 2013.

[106] Alessandro Gaggero, Francesco Martini, Francesco Mattioli, Fabio Chiarello, Robert Cernansky, Alberto Politi, and Roberto Leoni. Amplitude-Multiplexed readout of single photon detectors based on superconducting nanowires.

[107] R. HANBURY BROWN and R. Q. TWISS. Correlation between Photons in two Coherent Beams of Light. *Nature*, 177(4497):27, 1956.

[108] G. Reithmaier, M. Kaniber, F. Flassig, S. Lichtmannecker, K. Müller, A. Andrejew, J. Vučković, R. Gross, and J. J. Finley. On-Chip Generation, Routing, and Detection of Resonance Fluorescence. *Nano Letters*, 15(8):5208–5213, 2015.

[109] Svetlana Khasminskaya, Felix Pyatkov, Karolina Słowik, Simone Ferrari, Oliver Kahl, Vadim Kovalyuk, Patrik Rath, Andreas Vetter, Frank Hennrich, Manfred M. Kappes, Gregory Gol'tsmann, A. Korneev, Carsten Rockstuhl, Ralph Krupke, and Wolfram H. P. Pernice. Fully integrated quantum photonic circuit with an electrically driven light source. *Nature Photonics*, 10(11):727, 2016.

[110] Ulrich Rengstl. *III-V Semiconductor Photonic Integrated Circuits with Quantum Dots as Single-Photon Emitters*. Verlag Dr. Hut, 2017.

[111] Alan Migdall, Sergey V. Polyakov, Jingyun Fan, and Joshua C. Bienfang, editors. *Single-photon generation and detection*, volume 45 of *Experimental methods in the physical sciences*. Elsevier/Acad. Press, Amsterdam, 2013.

[112] Jacobus W. de Bakker. *Automata, languages and programming*, volume 85 of *Lecture notes in computer science*. Springer, Berlin, 1980.

[113] Adriano Barenco, Charles H. Bennett, Richard Cleve, David P. DiVincenzo, Norman Margolus, Peter Shor, Tycho Sleator, John A. Smolin, and Harald Weinfurter. Elementary gates for quantum computation. *Physical Review A*, 52(5):3457–3467, 1995.

[114] Colin P. Williams and Scott H. Clearwater. *Explorations in quantum computing*. Springer, New York, NY, 1998.

[115] Jozef Gruska. *Quantum computing*. Advanced topics in computer science series. McGraw-Hill, London, 1999.

[116] Michael A. Nielsen and Isaac L. Chuang. *Quantum computation and quantum information*. Cambridge Univ. Press, Cambridge, 10th anniversary ed. edition, 2010.

[117] T. C. Ralph, N. K. Langford, T. B. Bell, and A. G. White. Linear optical controlled-NOT gate in the coincidence basis. *Physical Review A*, 65(6):4337, 2002.

[118] T. B. Pittman, B. C. Jacobs, and J. D. Franson. Probabilistic quantum logic operations using polarizing beam splitters. *Physical Review A*, 64(6):46, 2001.

[119] C. L. Xu, W. P. Huang, M. S. Stern, and S. K. Chaudhuri. Full-vectorial mode calculations by finite difference method. *IEE Proceedings - Optoelectronics*, 141(5):281–286, 1994.

[120] C. J. M. Smith, H. Benisty, S. Olivier, M. Rattier, C. Weisbuch, T. F. Krauss, R. M. de La Rue, R. Houdré, and U. Oesterle. Low-loss channel waveguides with two-dimensional photonic crystal boundaries. *Applied Physics Letters*, 77(18):2813–2815, 2000.

[121] Stefan Hepp, Stephanie Bauer, Florian Hornung, Mario Schwartz, Simone L. Portalupi, Michael Jetter, and Peter Michler. Bragg grating cavities embedded into nano-photonic waveguides for Purcell enhanced quantum dot emission. *Optics express*, 26(23):30614–30622, 2018.

[122] Ivan N. Stranski and Lubomir Krastanow. Zur Theorie der orientierten Ausscheidung von Ionenkristallen aufeinander. *Monatshefte für Chemie und verwandte Teile anderer Wissenschaften*, 71(1):351–364, 1937.

[123] Peter Michler. *Single Quantum Dots*, volume 90 of *Topics in Applied Physics*. Springer, Berlin, 2003.

[124] Mario Schwartz. *Fully integrated GaAs-based quantum photonic circuits: resonant generation, splitting and detection of single-photon emission on-chip*. Physik. Dr. Hut, München, 2019.

[125] A. J. Bennett, D. C. Unitt, A. J. Shields, P. Atkinson, and D. A. Ritchie. Influence of exciton dynamics on the interference of two photons from a microcavity single-photon source. *Optics Express*, 13(20):7772, 2005.

[126] Simone Luca Portalupi, Gaston Hornecker, Valérian Giesz, Thomas Grange, Aristide Lemaître, Justin Demory, Isabelle Sagnes, Norberto D. Lanzillotti-Kimura, Loïc Lanco, Alexia Auffèves, and Pascale Senellart. Bright Phonon-Tuned Single-Photon Source. *Nano Letters*, 15(10):6290–6294, 2015.

[127] Hui Wang, Z-C Duan, Y-H Li, Si Chen, J-P Li, Y-M He, M-C Chen, Yu He, X. Ding, Cheng-Zhi Peng, Christian Schneider, Martin Kamp, Sven Höfling, Chao-Yang Lu, and Jian-Wei Pan. Near-Transform-Limited Single Photons from an Efficient Solid-State Quantum Emitter. *Physical Review Letters*, 116(21):213601, 2016.

[128] H. S. Nguyen, G. Sallen, C. Voisin, Ph. Roussignol, C. Diederichs, and G. Cassabois. Optically Gated Resonant Emission of Single Quantum Dots. *Phys. Rev. Lett.*, 108(5):057401, 2012.

[129] Alexander Högele, Stefan Seidl, Martin Kroner, Khaled Karrai, Richard J. Warburton, Brian D. Gerardot, and Pierre M. Petroff. Voltage-controlled optics of a quantum dot. *Physical Review Letters*, 93(21):217401, 2004.

[130] Chandra M. Natarajan, Michael G. Tanner, and Robert H. Hadfield. Superconducting nanowire single-photon detectors: physics and applications. *Superconductor Science and Technology*, 25(6):063001, 2012.

[131] A. Semenov, B. Günther, U. Böttger, H.-W. Hübers, H. Bartolf, A. Engel, A. Schilling, K. Ilin, M. Siegel, R. Schneider, D. Gerthsen, and N. A. Gippius. Optical and transport properties of ultrathin NbN films and nanostructures. *Physical Review B*, 80(5), 2009.

[132] Ekkehart Schmidt. *NbN SNSPD on GaAs for integrated photonic circuits.* Master Thesis, Karlsruher Institut für Technologie, Karlsruhe, 28.03.2014.

[133] A. Semenov, A. Engel, H.-W. Hübers, K. Il'in, and M. Siegel. Spectral cut-off in the efficiency of the resistive state formation caused by absorption of a single-photon in current-carrying superconducting nano-strips. *The European Physical Journal B*, 47(4):495–501, 2005.

[134] A. Engel, J. J. Renema, K. Il'in, and A. Semenov. Detection mechanism of superconducting nanowire single-photon detectors. *Superconductor Science and Technology*, 28(11):114003, 2015.

[135] M. Hofherr, D. Rall, K. Ilin, M. Siegel, A. Semenov, H.-W. Hübers, and N. A. Gippius. Intrinsic detection efficiency of superconducting nanowire single-photon detectors with different thicknesses. *Applied Physics Letters*, 108(1):014507, 2010.

[136] R. Lusche, A. Semenov, K. Il'in, Y. Korneeva, A. Trifonov, A. Korneev, H. Hubers, M. Siegel, and G. Gol'tsman. Effect of the Wire Width and Magnetic Field on the Intrinsic Detection Efficiency of Superconducting Nanowire Single-Photon Detectors. *IEEE Transactions on Applied Superconductivity*, 23(3):2200205, 2013.

[137] A. Engel, A. Aeschbacher, K. Inderbitzin, A. Schilling, K. Il'in, M. Hofherr, M. Siegel, A. Semenov, and H.-W. Hübers. Tantalum nitride superconducting single-photon detectors with low cut-off energy. *Applied Physics Letters*, 100(6):062601, 2012.

[138] A. Engel, K. Inderbitzin, A. Schilling, R. Lusche, A. Semenov, H. Hübers, D. Henrich, M. Hofherr, K. Il'in, and M. Siegel. Temperature-Dependence of Detection Efficiency in NbN and TaN SNSPD. *IEEE Transactions on Applied Superconductivity*, 23(3):2300505, 2013.

[139] D. Yu. Vodolazov. Single-Photon Detection by a Dirty Current-Carrying Superconducting Strip Based on the Kinetic-Equation Approach. *Physical Review Applied*, 7(3):1254, 2017.

[140] L. N. Bulaevskii, M. J. Graf, C. D. Batista, and V. G. Kogan. Vortex-induced dissipation in narrow current-biased thin-film superconducting strips. *Physical Review B*, 83(14):1915, 2011.

[141] F. Tafuri, J. R. Kirtley, D. Born, D. Stornaiuolo, P. G. Medaglia, P. Orgiani, G. Balestrino, and V. G. Kogan. Dissipation in ultra-thin current-carrying super-

conducting bridges; evidence for quantum tunneling of Pearl vortices. *Europhysics Letters (EPL)*, 73(6):948–954, 2006.

[142] H. Bartolf, A. Engel, A. Schilling, K. Il'in, M. Siegel, H.-W. Hübers, and A. Semenov. Current-assisted thermally activated flux liberation in ultrathin nanopatterned NbN superconducting meander structures. *Physical Review B*, 81(2):1140, 2010.

[143] John R. Clem and Karl K. Berggren. Geometry-dependent critical currents in superconducting nanocircuits. *Physical Review B*, 84(17):174510, 2011.

[144] Mohsen K. Akhlaghi, Haig Atikian, Amin Eftekharian, Marko Loncar, and A. Hamed Majedi. Reduced dark counts in optimized geometries for superconducting nanowire single photon detectors. *Optics Express*, 20(21):23610, 2012.

[145] M. Hofherr, D. Rall, K. Il'in, A. Semenov, H.-W. Hübers, and M. Siegel. Dark Count Suppression in Superconducting Nanowire Single Photon Detectors. *Journal of Low Temperature Physics*, 167(5-6):822–826, 2012.

[146] T. Yamashita, S. Miki, K. Makise, W. Qiu, H. Terai, M. Fujiwara, M. Sasaki, and Z. Wang. Origin of intrinsic dark count in superconducting nanowire single-photon detectors. *Applied Physics Letters*, 99(16):161105, 2011.

[147] Andrew J. Kerman, Eric A. Dauler, William E. Keicher, Joel K. W. Yang, Karl K. Berggren, G. Gol'tsman, and B. Voronov. Kinetic-inductance-limited reset time of superconducting nanowire photon counters. *Applied Physics Letters*, 88(11):111116, 2006.

[148] Alexei Semenov, Philipp Haas, Heinz-Wilhelm Hübers, Konstantin Ilin, Michael Siegel, Alexander Kirste, Dietemar Drung, Thomas Schurig, and Andreas Engel. Intrinsic quantum efficiency and electro-thermal model of a superconducting nanowire single-photon detector. *Journal of Modern Optics*, 56(2-3):345–351, 2009.

[149] Anthony J. Annunziata, Orlando Quaranta, Daniel F. Santavicca, Alessandro Casaburi, Luigi Frunzio, Mikkel Ejrnaes, Michael J. Rooks, Roberto Cristiano, Sergio Pagano, Aviad Frydman, and Daniel E. Prober. Reset dynamics and latching in niobium superconducting nanowire single-photon detectors. *Journal of Applied Physics*, 108(8):084507, 2010.

[150] Mariia Sidorova, Alexej Semenov, Heinz-Wilhelm Hübers, Ilya Charaev, Artem Kuzmin, Steffen Doerner, and Michael Siegel. Physical mechanisms of timing jitter in photon detection by current-carrying superconducting nanowires. *Phys. Rev. B*, 96(18):184504, 2017.

[151] Jason P. Allmaras, Alexander G. Kozorezov, Boris A. Korzh, Karl K. Berggren, and Matthew D. Shaw. Intrinsic timing jitter and latency in superconducting single photon nanowire detectors.

[152] D. Yu. Vodolazov. Minimal Timing Jitter in Superconducting Nanowire Single-Photon Detectors. *Physical Review Applied*, 11(1), 2019.

[153] Misael Caloz, Matthieu Perrenoud, Claire Autebert, Boris Korzh, Markus Weiss, Christian Schönenberger, Richard J. Warburton, Hugo Zbinden, and Félix Bussières. High-detection efficiency and low-timing jitter with amorphous superconducting nanowire single-photon detectors. *Applied Physics Letters*, 112(6):061103, 2018.

[154] Itamar Holzman and Yachin Ivry. Superconducting Nanowires for Single–Photon Detection: Progress, Challenges, and Opportunities. *Advanced Quantum Technologies*, 2(3-4):1800058, 2019.

[155] X. Zhang, A. Engel, Q. Wang, A. Schilling, A. Semenov, M. Sidorova, H.-W. Hübers, I. Charaev, K. Ilin, and M. Siegel. Characteristics of superconducting tungsten silicide WxSi1−x for single photon detection. *Physical Review B*, 94(17):239, 2016.

[156] Simone Ferrari, Vadim Kovalyuk, Wladislaw Hartmann, Andreas Vetter, Oliver Kahl, Changhyoup Lee, Alexander Korneev, Carsten Rockstuhl, Gregory Gol'tsman, and Wolfram Pernice. Hot-spot relaxation time current dependence in niobium nitride waveguide-integrated superconducting nanowire single-photon detectors. *Optics express*, 25(8):8739–8750, 2017.

[157] D. Henrich, P. Reichensperger, M. Hofherr, J. M. Meckbach, K. Il'in, M. Siegel, A. Semenov, A. Zotova, and D. Yu. Vodolazov. Geometry-induced reduction of the critical current in superconducting nanowires. *Phys. Rev. B*, 86(14):144504, 2012.

[158] Ilya Charaev, Alexey Semenov, Stefan Doerner, G. Gomard, Konstantin Ilin, and Michael Siegel. Current dependence of the hot-spot response spectrum of superconducting single-photon detectors with different layouts. *Superconductor Science and Technology*, 30(2):025016, 2016.

[159] F. B. Hagedorn and P. M. Hall. Right–Angle Bends in Thin Strip Conductors. *Journal of Applied Physics*, 34(1):128–133, 1963.

[160] H. L. Hortensius, E. F. C. Driessen, T. M. Klapwijk, K. K. Berggren, and J. R. Clem. Critical-current reduction in thin superconducting wires due to current crowding. *Applied Physics Letters*, 100(18):182602, 2012.

[161] Dagmar Henrich, L. Rehm, S. Dörner, M. Hofherr, K. Il'in, A. Semenov, and M. Siegel. Detection efficiency of a spiral-nanowire superconducting single-photon detector. *IEEE Transactions on Applied Superconductivity*, 23(3):2200405, 2013.

[162] W. H. P. Pernice, C. Schuck, O. Minaeva, M. Li, G. N. Goltsman, A. V. Sergienko, and H. X. Tang. High-speed and high-efficiency travelling wave single-photon detectors embedded in nanophotonic circuits. *Nature communications*, 3:1325, 2012.

[163] Patrik Rath, Oliver Kahl, Simone Ferrari, Fabian Sproll, Georgia Lewes-Malandrakis, Dietmar Brink, Konstantin Ilin, Michael Siegel, Christoph Nebel, and Wolfram Pernice. Superconducting single-photon detectors integrated with diamond nanophotonic circuits. *Light: Science & Applications*, 4(10):e338–e338, 2015.

[164] Mohsen K. Akhlaghi, Ellen Schelew, and Jeff F. Young. Waveguide integrated superconducting single-photon detectors implemented as near-perfect absorbers of coherent radiation. *Nature communications*, 6:8233, 2015.

[165] Andreas Vetter, Simone Ferrari, Patrik Rath, Rasoul Alaee, Oliver Kahl, Vadim Kovalyuk, Silvia Diewald, Gregory N. Goltsman, Alexander Korneev, Carsten Rockstuhl, and Wolfram H. P. Pernice. Cavity-Enhanced and Ultrafast Superconducting Single-Photon Detectors. *Nano Letters*, 16(11):7085–7092, 2016.

[166] Qimin Quan, Parag B. Deotare, and Marko Loncar. Photonic crystal nanobeam cavity strongly coupled to the feeding waveguide. *Applied Physics Letters*, 96(20):203102, 2010.

[167] Andrew Grieco, Boris Slutsky, Dawn T. H. Tan, Steve Zamek, Maziar P. Nezhad, and Yeshaiahu Fainman. Optical Bistability in a Silicon Waveguide Distributed Bragg Reflector Fabry–Pérot Resonator. *Journal of Lightwave Technology*, 30(14):2352–2355, 2012.

[168] F. Marsili, D. Bitauld, A. Gaggero, S. Jahanmirinejad, R. Leoni, F. Mattioli, and A. Fiore. Physics and application of photon number resolving detectors based on superconducting parallel nanowires. *New Journal of Physics*, 11(4):045022, 2009.

[169] Alessandro Gaggero, Francesco Mattioli, Zili Zhou, Rosalinda Gaudio, Roberto Leoni, and Andrea Fiore. Photon counting with a 24-pixel SSPD based photon number resolving detector. In *2016 18th International Conference on Transparent Optical Networks (ICTON)*, pages 1–4. IEEE, 10.07.2016 - 14.07.2016.

[170] M. J. Fitch, B. C. Jacobs, T. B. Pittman, and J. D. Franson. Photon-number resolution using time-multiplexed single-photon detectors. *Physical Review A*, 68(4):46, 2003.

[171] Tung-Sheng Yeh, Jenn-Ming Wu, and Wen-How Lan. The effect of AlN buffer layer on properties of AlxIn1−xN films on glass substrates. *Thin Solid Films*, 517(11):3204–3207, 2009.

[172] C. Stampfl and C. G. van de Walle. Density-functional calculations for III-V nitrides using the local-density approximation and the generalized gradient approximation. *Physical Review B*, 59(8):5521–5535, 1999.

[173] Glen A. Slack, R. A. Tanzilli, R. O. Pohl, and J. W. Vandersande. The intrinsic thermal conductivity of AIN. *Journal of Physics and Chemistry of Solids*, 48(7):641–647, 1987.

[174] N. Chaudhuri, R. S. Wadhwa, Phoola Tiku, and A. K. Sreedhar. Thermal Conductivity of Gallium Arsenide at Low Temperatures. *Physical Review B*, 8(10):4668–4670, 1973.

[175] Tatsuya Shiino, Shoichi Shiba, Nami Sakai, Tetsuya Yamakura, Ling Jiang, Yoshinori Uzawa, Hiroyuki Maezawa, and Satoshi Yamamoto. Improvement of the critical temperature of superconducting NbTiN and NbN thin films using the AlN buffer layer. *Superconductor Science and Technology*, 23(4):045004, 2010.

[176] P. M. Lundquist, W. P. Lin, Z. Y. Xu, G. K. Wong, E. D. Rippert, J. A. Helfrich, and J. B. Ketterson. Ultraviolet second harmonic generation in radio–frequency sputter–deposited aluminum nitride thin films. *Applied Physics Letters*, 65(9):1085–1087, 1994.

[177] M. Stegmaier, J. Ebert, J. M. Meckbach, K. Ilin, M. Siegel, and W. H. P. Pernice. Aluminum nitride nanophotonic circuits operating at ultraviolet wavelengths. *Applied Physics Letters*, 104(9):091108, 2014.

[178] D. Henrich, S. Dörner, M. Hofherr, K. Il'in, A. Semenov, E. Heintze, M. Scheffler, M. Dressel, and M. Siegel. Broadening of hot-spot response spectrum of superconducting NbN nanowire single-photon detector with reduced nitrogen content. *Journal of Applied Physics*, 112(7):074511, 2012.

[179] Dagmar Henrich. *Influence of Material and Geometry on the Performance of Superconducting Nanowire Single-Photon Detectors*, volume 010 of *Karlsruher Schriften-*

reihe zur Supraleitung. Technische Informationsbibliothek u. Universitätsbibliothek and KIT Scientific Publishing, Hannover and Karlsruhe, 2013.

[180] Mark A. Lewis, David A. Glocker, and Jacob Jorne. Measurements of secondary electron emission in reactive sputtering of aluminum and titanium nitride. *Journal of Vacuum Science & Technology A: Vacuum, Surfaces, and Films*, 7(3):1019, 1989.

[181] A. Nigro, G. Nobile, M. G. Rubino, and R. Vaglio. Electrical resistivity of polycrystalline niobium nitride films. *Physical Review B*, 37(8):3970–3972, 1988.

[182] Neil W. Ashcroft and N. David Mermin. *Solid state physics*. Brooks/Cole Thomson Learning, South Melbourne, repr edition, 1976.

[183] B. Abeles, Ping Sheng, M. D. Coutts, and Y. Arie. Structural and electrical properties of granular metal films. *Advances in Physics*, 24(3):407–461, 1975.

[184] S. Wolf and W. H. Lowrey. Zero Dimensionality and Josephson Coupling in Granular Niobium Nitride. *Physical Review Letters*, 39(16):1038–1041, 1977.

[185] A. Stockhausen, K. Il'in, M. Siegel, U. Södervall, P. Jedrasik, A. Semenov, and H-W Hübers. Adjustment of self-heating in long superconducting thin film NbN microbridges. *Superconductor Science and Technology*, 25(3):035012, 2012.

[186] John A. Thornton. The microstructure of sputter–deposited coatings. *Journal of Vacuum Science & Technology A: Vacuum, Surfaces, and Films*, 4(6):3059–3065, 1986.

[187] John A. Thornton. Influence of apparatus geometry and deposition conditions on the structure and topography of thick sputtered coatings. *Journal of Vacuum Science and Technology*, 11(4):666–670, 1974.

[188] Norbert Kaiser. Review of the fundamentals of thin-film growth. *Applied Optics*, 41(16):3053, 2002.

[189] Giulia Enrica Digeronimo. *Single-photon detectors integrated in quantum photonic circuits*. PHD thesis, Eindhoven University of Technology, Eindhoven, 2018.

[190] A. I. Gubin, K. S. Il'in, S. A. Vitusevich, M. Siegel, and N. Klein. Dependence of magnetic penetration depth on the thickness of superconducting Nb thin films. *Phys. Rev. B*, 72(6):224, 2005.

[191] Matthias Hofherr. *Real-time imaging systems for superconducting nanowire single-photon detector arrays*. Dissertation, Karlsruher Institut für Technologie, Karlsruhe, 2014.

[192] Edward Bennet Rosa. The self and mutual inductances of linear conductors. 1908, 1908.

[193] Anthony J. Annunziata, Daniel F. Santavicca, Luigi Frunzio, Gianluigi Catelani, Michael J. Rooks, Aviad Frydman, and Daniel E. Prober. Tunable superconducting nanoinductors. *Nanotechnology*, 21(44):445202, 2010.

[194] Keiji Yoshida, Mohammad Sajjad Hossain, Takanobu Kisu, Keiji Enpuku, and Kaoru Yamafuji. Modeling of Kinetic-Inductance Coplanar Striplin with NbN Thin Films. *Japanese Journal of Applied Physics*, 31(Part 1, No. 12A):3844–3850, 1992.

[195] Ch Kaiser, S. T. Skacel, S. Wünsch, R. Dolata, B. Mackrodt, A. Zorin, and M. Siegel. Measurement of dielectric losses in amorphous thin films at gigahertz frequencies using superconducting resonators. *Superconductor Science and Technology*, 23(7):075008, 2010.

[196] G. Hammer, S. Wuensch, M. Roesch, K. Ilin, E. Crocoll, and M. Siegel. Coupling of Microwave Resonators to Feed Lines. *IEEE Transactions on Applied Superconductivity*, 19(3):565–569, 2009.

[197] Tsuneo Konaka, Makoto Sato, Hidefumi Asano, and Shugo Kubo. Relative permittivity and dielectric loss tangent of substrate materials for high-T c superconducting film. *Journal of Superconductivity*, 4(4):283–288, 1991.

[198] Sonnet, 2011.

[199] Fabio Bernardini, Vincenzo Fiorentini, and David Vanderbilt. Polarization-Based Calculation of the Dielectric Tensor of Polar Crystals. *Physical Review Letters*, 79(20):3958–3961, 1997.

[200] Marc-Alexandre Dubois, Paul Muralt, and Laurent Sagalowicz. Aluminum nitride thin films for high frequency applications. *Ferroelectrics*, 224(1):243–250, 1999.

[201] Faraz Najafi, Jacob Mower, Nicholas C. Harris, Francesco Bellei, Andrew Dane, Catherine Lee, Xiaolong Hu, Prashanta Kharel, Francesco Marsili, Solomon Assefa, Karl K. Berggren, and Dirk Englund. On-chip detection of non-classical light by scalable integration of single-photon detectors. *Nature communications*, 6:5873, 2015.

[202] Mariia Sidorova, Alexej Semenov, Heinz-Wilhelm Hübers, Artem Kuzmin, Steffen Doerner, K. Ilin, Michael Siegel, Ilya Charaev, and Denis Vodolazov. Timing jitter in photon detection by straight superconducting nanowires: Effect of magnetic field and photon flux. *Physical Review B*, 98(13):445, 2018.

[203] P. Pfeffer and W. Zawadzki. Five-level $k \cdot p$ model for the conduction and valence bands of GaAs and InP. *Physical Review B*, 53(19):12813–12828, 1996.

[204] A. C. F. Hoole, M. E. Welland, and A. N. Broers. Negative PMMA as a high-resolution resist - the limits and possibilities. *Semiconductor Science and Technology*, 12(9):1166–1170, 1997.

[205] T. H. P. Chang. Proximity effect in electron–beam lithography. *Journal of Vacuum Science and Technology*, 12(6):1271–1275, 1975.

[206] M. Yu. Kupriyanov and V. F. Lukichev. Temperature dependence of pair-breaking current in superconductors. *Sov. J. Low Temp. Phys. (Engl. Transl.)*, 6(4):210–214, 1980.

[207] I. Charaev, T. Silbernagel, B. Bachowsky, A. Kuzmin, S. Doerner, K. Ilin, A. Semenov, D. Roditchev, D. Yu. Vodolazov, and M. Siegel. Proximity effect model of ultranarrow NbN strips. *Physical Review B*, 96(18):445, 2017.

[208] Ilya Charaev. *Improving the spectral bandwidth of superconducting nanowire single-photon detectors (SNSPDs)*. Dissertation, Karlsruher Institut für Technologie, Karlsruhe, 2017.

[209] K. Il'in, M. Hofherr, D. Rall, M. Siegel, A. Semenov, A. Engel, K. Inderbitzin, A. Aeschbacher, and A. Schilling. Ultra-thin TaN Films for Superconducting Nanowire Single-Photon Detectors. *Journal of Low Temperature Physics*, 167(5-6):809–814, 2012.

[210] J. C. Loredo, M. A. Broome, P. Hilaire, O. Gazzano, I. Sagnes, A. Lemaitre, M. P. Almeida, P. Senellart, and A. G. White. Boson Sampling with Single-Photon Fock States from a Bright Solid-State Source. *Physical Review Letters*, 118(13):130503, 2017.

[211] Alexandre Satrapinski, Ossi M. Hahtela, Alexander M. Savin, Sergey Novikov, and Natalia Lebedeva. Temperature Dependence of Pd Thin-Film Cryoresistors. *IEEE Transactions on Instrumentation and Measurement*, 60(7):2469–2474, 2011.

[212] Andrew J. Kerman, Joel K. W. Yang, Richard J. Molnar, Eric A. Dauler, and Karl K. Berggren. Electrothermal feedback in superconducting nanowire single-photon detectors. *Physical Review B*, 79(10), 2009.

[213] Junjie Wu, Lixing You, Sijing Chen, Hao Li, Yuhao He, Chaolin Lv, Zhen Wang, and Xiaoming Xie. Improving the timing jitter of a superconducting nanowire single-photon detection system. *Applied optics*, 56(8):2195–2200, 2017.

[214] Hüseyin Vural, Simone L. Portalupi, Julian Maisch, Simon Kern, Jonas H. Weber, Michael Jetter, Jörg Wrachtrup, Robert Löw, Ilja Gerhardt, and Peter Michler. Two-photon interference in an atom–quantum dot hybrid system. *Optica*, 5(4):367, 2018.

[215] Thomas Herzog. *Optical characterization of InGaAs quantum dots in GaAs/AlGaAs waveguide structures.* Master's thesis, University of Stuttgart, Stuttgart, 2015.

[216] Jonas Binz. *Optische untersuchungen von InAs/GaAs Quantenpunkten in integrierten Schaltkreisen.* Master's thesis, University of Stuttgart, Stuttgart, 2017.

[217] Florian Hornung. *On-chip Detection and Electrical Field Tuning of Quantum Dot Emission in Integrated GaAs Waveguide Circuits.* Master's thesis, University of Stuttgart, Stuttgart, 2018.

[218] L. Wang, A. Rastelli, and O. G. Schmidt. Structural and optical properties of In(Ga)As/GaAs quantum dots treated by partial capping and annealing. *Journal of Applied Physics*, 100(6):064313, 2006.

[219] Daniel Richter, Robert Hafenbrak, Klaus D. Jöns, Wolfgang-Michael Schulz, Marcus Eichfelder, Matthias Heldmaier, Robert Roßbach, Michael Jetter, and Peter Michler. Low density MOVPE grown InGaAs QDs exhibiting ultra-narrow single exciton linewidths. *Nanotechnology*, 21(12):125606, 2010.

[220] A. E. Grigorescu, M. C. van der Krogt, C. W. Hagen, and P. Kruit. 10nm lines and spaces written in HSQ, using electron beam lithography. *Microelectronic Engineering*, 84(5):822–824, 2007.

[221] Simone Luca Portalupi and Peter Michler. Resonantly Excited Quantum Dots: Superior Non-classical Light Sources for Quantum Information. In Peter Michler, editor, *Quantum Dots for Quantum Information Technologies*, volume 290 of *Nano-Optics and Nanophotonics*, pages 77–121. Springer International Publishing, Cham, 2017.

[222] Jin Liu, Kumarasiri Konthasinghe, Marcelo Davanço, John Lawall, Vikas Anant, Varun Verma, Richard Mirin, Sae Woo Nam, Jin Dong Song, Ben Ma, Ze Sheng Chen, Hai Qiao Ni, Zhi Chuan Niu, and Kartik Srinivasan. Single Self-Assembled InAs/GaAs Quantum Dots in Photonic Nanostructures: The Role of Nanofabrication. *Physical Review Applied*, 9(6), 2018.

[223] Ling Zhang, Lixing You, Dengkuan Liu, Weijun Zhang, Lu Zhang, Xiaoyu Liu, Junjie Wu, Yuhao He, Chaolin Lv, Zhen Wang, and Xiaoming Xie. Characterization of superconducting nanowire single-photon detector with artificial constrictions. *AIP Advances*, 4(6):067114, 2014.

[224] T. Flissikowski, A. Hundt, M. Lowisch, M. Rabe, and F. Henneberger. Photon Beats from a Single Semiconductor Quantum Dot. *Phys. Rev. Lett.*, 86(14):3172–3175, 2001.

[225] G. Sallen, A. Tribu, T. Aichele, R. André, L. Besombes, C. Bougerol, M. Richard, S. Tatarenko, K. Kheng, and J.-Ph. Poizat. Subnanosecond spectral diffusion measurement using photon correlation. *Nature Photonics*, 4:696 EP –, 2010.

[226] Michael Varnava, Daniel E. Browne, and Terry Rudolph. How good must single photon sources and detectors be for efficient linear optical quantum computation? *Physical Review Letters*, 100(6):060502, 2008.

List of Own Publications

Peer Reviewed Journals

[SIS17] E. Schmidt, K. Ilin, and M. Siegel. Aln-buffered superconducting nbn nanowire single-photon detector on gaas. *IEEE Transactions on Applied Superconductivity*, 27(4):1–5, 2017.

[SMW+13] Patrick M. Schwab, Carola Moosmann, Matthias D. Wissert, Ekkehart W-G Schmidt, Konstantin S. Ilin, Michael Siegel, Uli Lemmer, and Hans-Jürgen Eisler. Linear and nonlinear optical characterization of aluminum nanoantennas. *Nano Letters*, 13(4):1535–1540, 2013.

[SRS+19] Ekkehart Schmidt, Eric Reutter, Mario Schwartz, Huseyin Vural, Konstantin Ilin, Michael Jetter, Peter Michler, and Michael Siegel. Characterization of a photon-number resolving snspd using poissonian and sub-poissonian light. *IEEE Transactions on Applied Superconductivity*, page 1, 2019.

[SSR+18] Mario Schwartz, Ekkehart Schmidt, Ulrich Rengstl, Florian Hornung, Stefan Hepp, Simone L. Portalupi, Konstantin Llin, Michael Jetter, Michael Siegel, and Peter Michler. Fully on-chip single-photon hanbury-brown and twiss experiment on a monolithic semiconductor-superconductor platform. *Nano Letters*, 18(11):6892–6897, 2018.

Presentations on Conferences

[BSC+18] Reza Baghdadi, Ekkehart Schmidt, Ilya Charaev, Saman Jahani, Di Zhu, Michael Müller, Konstantin Ilin, Alexej Semenov, Jacob Zubin, Michael Siegel, and Karl K. Berggren. Varying thicknesses to minimize current-crowding in superconducting nanowire single photon detectors, 2018.

[RSS+17] Eric Reutter, Ekkehart Schmidt, Mario Schwartz, Hannes Rotzinger, Konstantin Ilin, Michael Jetter, Alexej Ustinov, and Michael Siegel. Superconducting nanowire single-photon detectors with multi-photon resolution for integrated quantum photonics on gaas, 2017.

[SCM+16] Ekkehart Schmidt, Ilya Charaev, Pavol Marko, J. M. Meckbach, Artem Kuzmin, Konstantin Ilin, and Michael Siegel. Ultra-thin nitride films for snspds on photonic substrates, 2016.

[SIS15] Ekkehart Schmidt, Konstantin Ilin, and Michael Siegel. Medium scale integration of superconducting nbn nanowire single-photon detectors on gaas for quantum photonic integrated circuits, 2015.

[SKS+13] Ekkehart Schmidt, Elisabeth Koroknay, Mark Sartison, Konstantin Ilin, Michael Jetter, Peter Michler, and Siegel Michael. Nbn snspd on gaas for photonic integrated circuits, 2013.

[SMD+14] Patrick M. Schwab, Carola Moosmann, Katja Dopf, Matthias D. Wissert, Ekkehart Schmidt, Konstantin Ilin, Michael Siegel, Uli Lemmer, and Hans-Jürgen Eisler. Aluminium resonant optical antenna resonance spectra from two-photon laser excitation, 2014.

[SMIS15] Ekkehart Schmidt, Michael Merker, Konstantin Ilin, and Michael Siegel. Superconducting nbn single-photon detectors on gaas with an aln buffer layer, 2015.

[SRK+14a] Ekkehart Schmidt, Ulrich Rengstl, Elisabeth Koroknay, Konstantin Ilin, Michael Jetter, Peter Michler, and Siegel Michael. Nbn snspds on gaas for integrated photonic circuits, 2014.

[SRK+14b] Ekkehart Schmidt, Ulrich Rengstl, Elisabeth Koroknay, Konstantin Ilin, Michael Jetter, Peter Michler, and Siegel Michael. Thin nbn film nanowires on gaas for single-photon detectors, 2014.

[SRR+17] Ekkehart Schmidt, Eric-René Reutter, Hannes Rotzinger, Konstantin Ilin, Alexej Ustinov, and Michael Siegel. Photon number resolving superconducting nanowire single-photon detectors, 2017.

[SRS+17a] Ekkehart Schmidt, Eric-René Reutter, Mario Schwartz, Hannes Rotzinger, Konstantin Ilin, Michael Jetter, Peter Michler, Alexej Ustinov, and Siegel Michael. Photon-number-resolving detectors operating at high repetition rates, 2017.

[SRS+17b] Ekkehart Schmidt, Eric-René Reutter, Mario Schwartz, Hannes Rotzinger, Konstantin Ilin, Michael Jetter, Peter Michler, Alexej Ustinov, and Siegel Michael. Ultrafast photon-number-resolving detectors for nir operating at 4.2 k, 2017.

[SRS+18] Ekkehart Schmidt, Eric-René Reutter, Mario Schwartz, Konstantin Ilin, Alexej Ustinov, and Michael Siegel. Photon-number resolving snspd for integrated photonics, 2018.

[SSH+16a] Ekkehart Schmidt, Mario Schwartz, Thomas Herzog, Konstantin Ilin, Michael Jetter, Peter Michler, and Michael Siegel. Influence of back reflections on the detection efficiency of superconducting nanowire single-photon detectors on gaas, 2016.

[SSH+16b] Ekkehart Schmidt, Mario Schwartz, Thomas Herzog, Konstantin Ilin, Michael Jetter, Peter Michler, and Michael Siegel. Influence of parasitic counts on the detection efficiency of superconducting nanowire single-photon detectors on gaas, 2016.

[SSH+18] Ekkehart Schmidt, Mario Schwartz, Florian Hornung, Ulrich Rengstl, Stefan Hepp, Konstantin Ilin, Simone Luca Portalupi, Michael Jetter, Michael Jetter, and Michael Siegel. Superconducting nanowire single-photon detectors on gaas with suppressed parasitic counts, 2018.

[SSR+16] Ekkehart Schmidt, Mario Schwartz, Ulrich Rengstl, Thomas Herzog, Konstantin Ilin, Michael Jetter, Peter Michler, and M. Siegel. Suppression of straylight counts of snspds integrated into integrated quantum-photonic circuits, 2016.

[SSR+18] Ekkehart Schmidt, Mario Schwartz, Ulrich Rengstl, Florian Hornung, Stefan Hepp, Simone L. Portalupi, Konstantin Ilin, Michael Jetter, and Peter Michler. Fully on-chip single-photon hanbury-brown and twiss experiment on a monolithic semiconductor–superconductor platform, 2018.

Supervised Student Theses

[Bar17] Spiro Barjami. *Optimization of AlN/NbN bilayers for SNSPD on GaAs*. Bachelor's thesis, Karlsruher Institut für Technologie, Karlsruhe, 2017.

[Ben15] Hannes Bender. *Optimierung von AlN-Pufferschichten für supraleitende Einzelphotonendetektoren*. Bachelor's thesis, Karlsruher Institut für Technologie, Karlsruhe, 2015.

[Bra17] Nicholas-Philip Brandt. *Automatisierung eines optischen Messsystems zur positionsabhängigen Characterisierung supraleitender Detektoren*. Bachelor's thesis, Karlsruher Institut für Technologie, Karlsruhe, 2017.

[Mör16] Daniel Mörixbauer. *Aufbau eines Dünnschicht NbN/AlN/NbN Hochpasses für supraleitende Hochfrequenzanwendungen*. Bachelor's thesis, Karlsruher Institut für Technologie, Karlsruhe, 2016.

[Mül18] Michael Müller. *Supraleitender Einzelphotonendetektor mir variirender Dicke*. Bachelor thesis, Karlsruher Institut für Technologie, Karlsruhe, 2018.

[Reu17] Eric-René Reutter. *Photon-number resolving superconducting nanowire single-photon detectors for quantum photonics*. Master's thesis, Karlsruher Institut für Technologie, Karlsruhe, 2017.

Danksagung

Das Schreiben einer Doktorarbeit ist nur möglich durch die Unterstützung von verschiedenen Seiten. An dieser Stelle möchte ich mich bei allen bedanken, die mich in den letzten Jahren begleitet und unterstützt haben.
An erster Stelle möchte ich mich bei meinem Doktorvater Michael Siegel bedanken, der mir ermöglicht hat mit großer Freiheit an einem hochinteressanten Projekt zu Arbeiten und die erreichten Ergebnisse auf einer Vielzahl wissenschaftlicher Tagungen zu präsentieren. Vielen Dank für das Vertrauen, das Sie in mich gesetzt haben. Bei *Konstantin Ilin* möchte mich herzlich für eine seit meiner Bachelorarbeit andauernde Betreuung bedanken. Durch seine große Erfahrung konnte er mir bei auftretenden wissenschaftlichen und technischen Herausforderungen weiterhelfen und war immer ein guter Ratgeber. Seine Ideen und Vorschläge waren eine große Hilfe. Vielen Dank an *Stefan Wünsch* für die Hilfe bei administrativen Fragen und bei Fragen zu Hochfrequenzmessungen und -design. Bei meinen Kollegen: *Artem Kuzmin, Michael Merker, Steffen Dörner, Ilya Charaev, Alexander Schmid, Juliane Raasch* und *Matthias Arndt* möchte ich mich für die Zusammenarbeit, die fachlichen Diskussionen und die Unterstützung in technischen Fragen bedanken. Vielen Dank an meine Bachelorstudenten, die meine Arbeit durch tatkräftige Mitarbeit unterstützt haben (in chronologischer Reihenfolge): *Hannes Bender, Daniel Mörixbauer, Nicholas-Philipp Brandt, Spiro Barjami* und *Michael Müller.* Ein besonderer Dank hierbei gilt meinem Masterstudenten *Eric-René Reutter* der mich nicht nur während seiner Masterarbeit unterstützt hat, sondern darüber hinaus als Doktorand am Max-Planck-Institut für Festkörperforschung in Stuttgart Wellenleiter für mich belichtet hat. Vielen Dank an *Alexander Stassen* und *Karlheinz Gutbrod* für die häufige technische Unterstützung und an *Doris Duffner*, die einem bei administrativen Dingen den Rücken freihält. Danke an *Frank Ruhnau* für eine schnelle IT Unterstützung bei auftretenden Problemen.
Diese Arbeit wurde in enger Zusammenarbeit mit dem Institut für Halbleiteroptik und funktionelle Grenzflächen (IHFG) der Universität Stuttgart durchgeführt. Hier möchte ich mich beim Institutsleiter *Peter Michler* bedanken, der es versteht einen zu Bestleistungen zu motivieren. Vielen Dank an *Michael Jetter* und *Simone Portalupi* für die vielen Gespräche und fachliche Diskussionen, die das PIC-Projekt vorangebracht haben. Besonders bedanken möchte ich mich bei *Ulrich Rengstl* und *Mario Schwartz* (*Waveguys*). Nur durch die enge und hervorragende Zusammenarbeit mit euch war es möglich einen vollintegrierten Chip zu realisieren. *Ulrich Rengstl* für das Simulieren, Entwerfen und Ätzen der besten Wellenleiterstrukturen und *Mario Schwartz* (ja, mit tz) für die vielen optischen Messungen mit und ohne Quantenpunktanregung unterstützt von *Thomas Herzog, Jonas Binz* und *Florian Hornung*. Vielen Dank *Mario* für die unzähligen gemeinsamen Messtage, bei denen ich bei dir übernachten durfte und für deine Geduld, wenn mal nicht alles perfekt lief. Vielen Dank an *Stefan Hepp* für das Wachsen von Quantenpunktproben, das Ätzen der letzten Wellenleiter Proben und an *Stephanie Bauer* für das Belichten und Ätzen von Wellenleiter Strukturen

für diverse Liftofftests. Vielen Dank an *Thomas Reindl* vom Max-Planck-Institut für Festkörperforschung in Stuttgart für das Belichten von Wellenleiterstrukturen. Neben den bereits genannten, möchte ich mich bei allen anderen bedanken, die mir am IHFG das Gefühl gegeben haben nicht nur Kooperationspartner, sondern fester Teil des Instituts zu sein: *Marc Sartison, Hüseyin Vural, Roman Bek, Norbert Witz, Simon Seyfferle, Cornelius Nawrath, Fabian Olbrich....* Vielen Dank an meinen langjährigen Kletterpartner *Johannes*, für viele Kletterabende und das Probelesen meiner Doktorarbeit. Danke an mein Quizteam *"Die Ofenkartoffeln"* für eine tolle Zeit und viele erfolgreiche Quizabende. Und an alle Freunde, die meine Zeit in Karlsruhe begleitet haben und mir immer eine Gelegenheit dazu gegeben haben auch mal abschalten zu können.

Ein großer Dank gilt meiner Familie, besonders den beiden besten Schwestern die man haben kann, dafür das sie mich immer unterstützen und an mich glauben. Und meiner liebsten *Ann-Kathrin*, die mir den Rücken gestärkt hat und im letzten Jahr zu oft auf mich verzichten musste. Ich freue mich auf unsere Gemeinsame Zukunft.

Karlsruher Schriftenreihe zur Supraleitung

Karlsruher Institut für Technologie (KIT) | ISSN 1869-1765

Herausgeber: Prof. Dr.-Ing. M. Noe, Prof. Dr. rer. nat. M. Siegel

Band 001 **Christian Schacherer**
Theoretische und experimentelle Untersuchungen zur Entwicklung supraleitender resistiver Strombegrenzer. 2009
ISBN 978-3-86644-412-6

Band 002 **Alexander Winkler**
Transient behaviour of ITER poloidal field coils. 2011
ISBN 978-3-86644-595-6

Band 003 **André Berger**
Entwicklung supraleitender, strombegrenzender Transformatoren. 2011
ISBN 978-3-86644-637-3

Band 004 **Christoph Kaiser**
High quality Nb/Al-AlOx/Nb Josephson junctions. Technological development and macroscopic quantum experiments. 2011
ISBN 978-3-86644-651-9

Band 005 **Gerd Hammer**
Untersuchung der Eigenschaften von planaren Mikrowellenresonatoren für Kinetic-Inductance Detektoren bei 4,2 K. 2011
ISBN 978-3-86644-715-8

Band 006 **Olaf Mäder**
Simulationen und Experimente zum Stabilitätsverhalten von HTSL-Bandleitern. 2012
ISBN 978-3-86644-868-1

Band 007 **Christian Barth**
High Temperature Superconductor Cable Concepts
for Fusion Magnets. 2013
ISBN 978-3-7315-0065-0

Band 008 **Axel Stockhausen**
Optimization of Hot-Electron Bolometers for THz Radiation. 2013
ISBN 978-3-7315-0066-7

Band 009 **Petra Thoma**
Ultra-fast $YBa_2Cu_3O_{7-x}$ direct detectors for the THz
frequency range. 2013
ISBN 978-3-7315-0070-4

Band 010 **Dagmar Henrich**
Influence of Material and Geometry on the Performance
of Superconducting Nanowire Single-Photon Detectors. 2013
ISBN 978-3-7315-0092-6

Band 011 **Alexander Scheuring**
Ultrabreitbandige Strahlungseinkopplung in THz-Detektoren. 2013
ISBN 978-3-7315-0102-2

Band 012 **Markus Rösch**
Development of lumped element kinetic inductance detectors
for mm-wave astronomy at the IRAM 30 m telescope. 2013
ISBN 978-3-7315-0110-7

Band 013 **Johannes Maximilian Meckbach**
Superconducting Multilayer Technology for Josephson
Devices. 2013
ISBN 978-3-7315-0122-0

Band 014 **Enrico Rizzo**
Simulations for the optimization of High Temperature
Superconductor current leads for nuclear fusion applications. 2014
ISBN 978-3-7315-0132-9

Band 015 **Philipp Krüger**
Optimisation of hysteretic losses in high-temperature superconducting wires. 2014
ISBN 978-3-7315-0185-5

Band 016 **Matthias Hofherr**
Real-time imaging systems for superconducting nanowire single-photon detector arrays. 2014
ISBN 978-3-7315-0229-6

Band 017 **Oliver Näckel**
Development of an Air Coil Superconducting Fault Current Limiter. 2016
ISBN 978-3-7315-0526-6

Band 018 **Christoph M. Bayer**
Characterization of High Temperature Superconductor Cables for Magnet Toroidal Field Coils of the DEMO Fusion Power Plant. 2017
ISBN 978-3-7315-0605-8

Band 019 **Shengnan Zou**
Magnetization of High Temperature Superconducting Trapped-Field Magnets. 2017
ISBN 978-3-7315-0715-4

Band 020 **Ilya Charaev**
Improving the Spectral Bandwidth of Superconducting Nanowire Single-Photon Detectors (SNSPDs). 2018
ISBN 978-3-7315-0745-1

Band 021 **Juliane Raasch**
Electrical-field sensitive $YBa_2Cu_3O_{7-x}$ detectors for real-time monitoring of picosecond THz pulses. 2019
ISBN 978-3-7315-0786-4

Band 022 **Yingzhen Liu**
Design of a superconducting DC wind generator. 2019
ISBN 978-3-7315-0796-3

Band 023	**Sebastian Hellmann**
Research and Technology Development on Superconducting Current Limiting Transformers. 2019
ISBN 978-3-7315-0804-5

Band 024	**Simon J. Otten**
Characterisation of REBCO Roebel cables. 2019
ISBN 978-3-7315-0904-2

Band 025	**Julia Brandel**
Supraleitende Einzelphotonenzähler: Optimierung der Zeitauflösung und Anwendungsbeispiele aus der Spektroskopie. 2019
ISBN 978-3-7315-0917-2

Band 026	**Dustin Kottonau, Eugen Shabagin, Wescley T. B. de Sousa, Jörn Geisbüsch, Mathias Noe, Hanno Stagge, Simon Fechner, Hannes Woiton, Thomas Küsters**
Bewertung des Einsatzes supraleitender 380-kV-Kabel. 2019
ISBN 978-3-7315-0927-1

Band 027	**Steffen Dörner**
Multifrequenzausleseverfahren von supraleitenden Einzelphotonen-Detektoren. 2019
ISBN 978-3-7315-0961-5

Band 028	**Michael Merker**
Superconducting integrated THz receiver. 2019
ISBN 978-3-7315-0970-7

Band 029	**Wolfgang-Gustav Ekkehart Schmidt**
Superconducting Nanowire Single-Photon Detectors for Quantum Photonic Integrated Circuits on GaAs. 2020
ISBN 978-3-7315-0980-6